Restructured Power System and Electricity Market Forecasting

First Edition

Edited by

M. M. Tripathi

PREFACE

With the introduction of deregulation in power industry, many challenges have been faced by the participants in the emerging electricity market. The fundamental objective of electric power industry deregulation is efficient generation, consumption of electricity, and reduction in energy prices. In India the Electricity Act 2003 initiated a liberal framework for development of the power industry, promoting competition, protecting interests of consumers and supply of electricity to all areas, rationalization of electricity tariff and ensure transparent policies and promotion of efficiency in Indian Power System. With Open access, the market structure in the Indian Power sector started changing from the old single buyer structure to a multi-buyer market model. The Indian power sector is now in the process of being restructured towards free market based system. The success of the new deregulated power system completely depends upon creation of competition in power market. To achieve this there may be different ways of restructuring the power industry but considering the organizational set-up, financial condition, control structure and their coordination many models are in place or proposed for implementing it gradually as the system gets matured in India. A good restructured power system with high level of competition would provide cheaper and good quality electricity to the consumer and also increase the profit of all market entities.

To achieve these goals, accurate and efficient electricity load, price and ancillary services forecasting has become more important. Accurate forecasting of electricity demand not only will help in optimizing the start-up of generating units it also save the investment in the construction of required number of power facilities and help to check the risky operation and unmet demand, demand of spinning reserve, and vulnerability to failures. Price forecasting provide crucial information for power producers and consumers to develop bidding strategies in order to maximize profit. It plays an important role in power system planning and operation, risk assessment and other decision making. Its main objective is to reduce the cost of electricity through competition, and maximize efficient generation and consumption of electricity. Because of the non-storable nature of electricity, all generated electricity must be consumed. Therefore, both producers and consumers need accurate price forecasts in order to establish their own strategies for benefit or utility maximization. Many factors which influence the electricity price, such as hour of the day, day of the week, month, year, historical prices and demand, natural gas price etc.

This book presents the overview of present power system structure and capacity in India. In this book various issues related with power system restructuring has been covered in detail. Also various pricing mechanism and market entities as well as structure have been presented. The issues related to the operation of electricity market have been discussed. Importance of forecasting of electricity market parameters has been discussed. Neural Network based design of forecasting tool to compute the short-term load, price and spinning reserve forecast of different power market has been discussed to show how such method can be useful for operation, security and reliability of the electricity market.

The content of this book would be very useful to undergraduate and post graduate students of Electrical Engineering. Tis book will also help researchers working in the area of electricity market forecasting to develop an insight of the problem and initiate them to find solutions of some of the problems of power system restructuring.

M. M. Tripathi

ABOUT THE AUTHOR

Madan Mohan Tripathi is Associate Professor in Delhi Technological University, New Delhi, India. He has done his B. E. (Electrical Engineering) from Madan Mohan Malviya Engineering College, Gorakhpur, India and Ph. D. (Electrical Engineering) from G. B. Technical University, Lucknow, India. He has worked with Institute for Plasma Resaerch, Gandhinagar, India and National Institute of Electronics & Information Technology (NIELIT), Delhi, India. His research interests include sustainable energy, power system restructuring, AI application to power system problems and application of IT in power system control and monitoring. He is a member of Plasma Science Society of India (PSSI), IEEE and IETE.

INDEX

CHAPTER 1

Restructuring of Power System in India

1.1 Overview of Indian Power Sector

India is the fifth largest generator and consumer of electric power in the world. It has installed capacity of more than 200000 MW as shown in table 1.1 below. The main power generation in India is due to thermal, hydro and nuclear generation. With the new policies of government of India, contribution of renewable energy sources has increased in recent years. About 60% of the generation capacity was owned by private sector when India got independence but the situation is entirely different today and approx. 70% of the installed capacity belongs to public sector. Also the ratio of hydro to thermal power generation capacity is reducing day by day. At present Indian power sector has approx. 68% thermal, 18% hydro, 12% renewable & 2% nuclear energy share of total generation capacity [1].

Table 1.1: Installed Generation Capacity (MW) as on 31.03.2013

Source	Central	State	Private	Total Installed Capacity
Thermal	51121	57939	42471	151530
Coal	44055	51661	34505	130221
Hydro	9459	27437	2595	39491
Renewable Energy Sources (RES)	0	3748	23794	27542
Gas	7066	5676	7368	20110
Nuclear	4780	0	0	4780
Diesel	0	603	597	1200
Total Installed Capacity	65,360	89,125	68,859	2,23,344

Source: http://www.cea.nic.in/

India has seen tremendous growth in terms of power generation capacity since Independence. The total installed power generation capacity at the end of first plan was approx. 3000 MW, which has reached to around 200000 MW in April 2012. The plan-wise growth in installed power generation capacity in India is presented in table 1.2 [2].

Table 1.2: Plan-wise Growth in Installed Generation Capacity in India

Particulars	Installed Power Generation Capacity (MW)
First Five Year Plan	2886
Second Five Year Plan	4653
Third Five Year Plan	9027
Fourth Five Year Plan	16664
Fifth Five Year Plan	26680
Sixth Five Year Plan	42585
Seventh Five Year Plan	63636
Eighth Five Year Plan	85795
Ninth Five Year Plan	105046
Tenth Five Year Plan	132329
Eleventh Five Year Plan	199877

Source: http://www.cea.nic.in/

Table 1.3: Plan-wise Growth in Indian Transmission Sector

Particulars	400 KV		220 KV	
	Transmission Lines (Ckt-Km)	Sub-station (MVA)	Transmission Lines (Ckt-Km)	Sub-station (MVA)
Sixth Five Year Plan	6029	9330	46005	37291
Seventh Five Year Plan	19824	21580	59631	53742
Eighth Five Year Plan	36142	40865	79600	84177
Ninth Five Year Plan	49378	60380	96993	116363
Tenth Five Year Plan	75722	92942	114629	156497
Eleventh Five Year Plan	113367	150127	140164	223774

Source: CEA

After Independence India has seen a gradual increase in the transmission voltage levels starting at 132 kV to 220 kV and now at 400 KV [3]. The power transmission in India has witnessed a massive transformation and up-gradation as shown in table 1.3. Power Sector institutional framework and list of entities in India is shown in table 1.4 and main power sector organizations in India are listed in table 1.5 below.

Table 1.4: Power Sector Institutional Framework & Entity Responsibility

Entity	Responsibilities
Ministry of Power (MoP)	Lays down Acts & Legal provisions, policies and guidelines for competitive bidding.
Central Electricity Authority (CEA)	Formulates national electricity plan, monitors projects, and maintains data and statistics and forecast the demand.
Regulatory Commission	Regulates Power sector entities, decides tariffs, monitors supply and service quality and ensures implementation of provisions of Acts related to electricity.
Generator	Generates Power based on contracts or independently. Follows load dispatch directions for scheduling the generation.
Transmission Owner	Builds, operates and maintains the transmission network infrastructure.
Distributor	Builds, operates and maintains the distribution network, supply electricity to consumers and metering, billing and money collection from consumers.
Trader	Facilitates transaction of Power at negotiable prices in the power network.
Load Dispatch Centre (LDC)	Maintains grid stability and discipline. A statutory body entrusted with responsibility of scheduling and accounting the Power.

Source: http://indiasmartgrid.org

Central government of India formed the large organizations like National Thermal Power Corporation (NTPC), National Hydro Power Corporation (NHPC) and Nuclear Power Corporation (NPC) to install the large power generation capacities in India. In addition to this state governments have also installed their own generation capacities. While we have augmented the power generation, transmission and distribution in a great way, the present Indian power sector is also facing many problems such as revenue shortage due to low tariff, electricity subsidy, system losses, loss in revenue collection, insufficient budget provisions, lack of investment from private sector and lack of awareness among public [1], [4].

Table 1.5: Prominent Organizations in Indian Power Sector

Organization	Expertise	Established
Damodar Valley Corporation Ltd.	Generation & Distribution of Power in Damodar Valley region.	1948
Central Power Research Institute (CPRI)	Research in Power engineering, testing and certification of Power equipment.	1960
National Power Training Institute (NPTI)	National Apex body for Training and Human Resources Development in Power Sector.	1965
Bhakra Beas management Board (BBMB)	Administration, Operation and Maintenance of Hydro Power projects.	1967
Rural Electrification Corporation (REC)	Financial assistance to SEBs, State Governments and Rural Cooperatives for Rural electrification projects.	1969
National Thermal Power Corporation (NTPC)	Design, commissioning and operation of Thermal Power Plants. Centrally owned.	1975
National Hydro Electric Power Corporation (NHPC	Design, commissioning and operation of Hydro Power Plants. Centrally owned.	1975
North Eastern Electric Power Corporation (NEEPCO)	Design, commissioning and operation of Power projects in North Eastern region.	1976
Power Finance Corporation (PFC)	Power Sector financing of the Power and associated sectors.	1986
Satluj Jal Vidyut Nigam (SJVN)	Operating, commissioning and maintaining Hydropower projects.	1988
Power Grid Corporation of India Ltd. (PGCIL)	Central Transmission utility involved in establishment, operation and maintenance of inter-regional grids.	1989
Power Trading Corporation (PTC)	Development of power market and electricity trading in India.	1999
Bureau of Energy Efficiency (BEE)	Improvement of energy efficiency through regulatory and promotional mechanism.	2002

Source: http://powermin.nic.in/

The figures for accumulated losses are mind boggling and thought provocative. As per the Power Finance Corporation (PFC) Report on "Performance of State Power Utilities for FY 2008-09 to FY 2010-11", the accumulated loss of power utilities has already surpassed the Rs. 1.16 trillion figure in March 2011 [5]. This certainly calls for greater autonomy and regulatory institutional reforms.

1.2 Deregulation of Power Sector World-wide

The restructuring in the developed countries has forced other countries to take suitable steps for lower price, better quality, improved reliability and higher efficiency of the system. The forces behind worldwide electric sector deregulation have either been political reform, regulatory failures, high tariffs, managerial inefficiency or global economic crisis [6]. The reasons for deregulation are different in different countries. Many countries made the changes as a result of the failure of the state to adequately manage electricity companies while in other countries lack of public resources to finance the required investment for the development was the reason of deregulation and World Bank loans to the utilities was a major incentive to start the deregulation process [7].

The traditional power industry has one or many of the characteristics listed below.

- The government owned national or state electric utility was permitted to produce, transmit, distribute and sell commercial electric power.
- The utilities had to provide electricity for all consumers without the consideration of profit and it was obligatory to serve.
- The government regulator lay down the guidelines and rules for utility's business and operating practices as well as and rate of electricity which was mandatory to follow.
- The profit margin of the utilities above its cost was ensured by the government in terms of financial support.
- The electric utilities were directed to operate in a way to minimize the overall revenue requirements.

For many years, it has been assumed that electricity and its delivery were inseparable but now electricity markets are open to alternative producers and alternative purchasers. Efforts are underway to sell electricity as a product that can be bought and transported from place to place. It provided the drive for the newly emerging world of competition in production and choice for consumers [8].

A wide variety of efforts and experiences are going on in reorganizing the electricity supply industry around the world by attempting to combine two inherently different regulatory approaches of the industry i.e. traditional cost-of-service regulator and fully competitive markets. Governments and regulators around the world are considering whether they should restructure and/or privatize their electric supply industry [9]. Restructuring and privatization are the different dimensions of change in electricity sector. Restructuring is the commercial arrangements for selling energy, separating or unbundling integrated industry structures and introducing competition and choice. Privatization is a change in ownership from Government to Private management [10]. Motive behind deregulation is to increase efficiency through better investment decisions and management, optimized use of existing plants, choice for customers. The reason may be any one or many but their alternatives and the implications must be known before implementation.

1.2.1 *Deregulation in Indian Power Sector*

Electricity sector is in India involve both central and state governments. The Ministry of Power has overall authority for power sector development [11]. During the last decade, electric supply industry in India has been continuously changing from the monopoly structure to a competitive market. The restructuring of the Indian power sector started due to the scarcity of financial resources available with Central and State Governments and necessity of improving the operational and commercial efficiency [12].

Indian Electricity Act in the Year 1887 aimed to regulate the Generation, Supply and use of electricity and provided protection of the person as well as property, from injury and risks arising from the supply and usage of electricity for lighting and other purposes. This Act was repealed and replaced by the Indian Electricity Act, 1903 (III of 1903). Many practical, electro technical and commercial difficulties were realized during the period of 1903 to 1909. The Indian Electricity Bill was passed by the Legislative Council on 18th

March, 1910 and it became the Indian Electricity Act, 1910 (9 of 1910) with effect from 1st January, 1911. In 1948, the Government of India enacted the Electricity Supply Act to make the way for the formation of the State Electricity Boards. These State Electricity Boards were supposed to develop networks of transmission lines and add generation capacity. The milestones legislations in Indian power sector are presented in table 1.6 below.

Table 1.6: Milestone Legislations in Indian Power Sector

Laws/Policies	Objective	Impact
The Electricity Act, 1948	Mandated creation of SEBs	Ownership in the hands of SEBs
The Regulatory Commission Act, 1998	Provision of setting-up of Central/State Electricity Regulatory Commission	Independent Regulatory mechanism
Electricity Act, 2003	Providing reliable & quality power to Customers at reasonable rate	Investments in Capacity addition

Source: http://www.doe.gov

Based on the Government's regulation and policies, the evolution of Regulatory framework can be divided into two board phases:

(i) The First Phase- from 1948 to 1997 was marked by constitution of Central Electricity Authority (CEA) and growth in Generation & Transmission capacities through the establishment of State Electricity Boards & Central Public sector Units.

(ii) The Second phase is marked by the constitution of Regulatory Commissions in 1948 and the reforms post the introduction of Electricity Act, 2003.

Under the Electricity (Supply) Act 1948, the Central Electricity Authority (CEA) was established at Central level and the State Electricity Boards (SEBs) at the state level. The objective of the CEA was to develop a uniform National Power Policy to for development of the Power sector in India. The SEBs took over the private companies in their respective states. While the SEBs aided the growth in the Indian electricity sector, but later on SEBs suffered huge financial and technical losses. Government was forced to restructure the sector and take appropriate Regulatory initiatives to distance the Tariff setting process from State Government.

Till 1990, the power sector in India was evolved as a public monopoly. With an aim to distance the Government from tariff setting, Government of India enacted the Electricity Regulatory Commissions Act 1998 which was aimed towards creation of independent regulators in the energy sector. This paved the way for setting up of the Central Electricity Regulatory Commission (CERC) and State Electricity Regulatory Commissions (SERCs). The functions of the CERC and the SERCs are clearly demarcated. While the CERC are responsible for all centrally owned power stations and other stations having an inter-state role, the SERCs are responsible for stations within their own state only [13]. Through Electricity Regulatory Commissions Act 1998, Government had liberalized the energy sector and opened the sector for foreign and private investments to increase the availability of funds for the Power sector. Ministry of Power started Accelerated Power Development and Reform Programme (APDRP) from the year 2000-01 with the objectives of making financial turn-around in the performance of the power sector especially in electric distribution and improvement in quality of supply [14].

These reforms ensured that power sector is financially viable and attractive enough for private investors to investment in energy sector and tariffs were determined according to economic principles and the entire tariff determinative process becomes free from any political interference. Now the Government is a facilitator which would lay down broad principles of policy and monitors its implementation.

On June 10, 2003 the electricity Act 2003 came into effect which took place of all the earlier laws and acts such as the Indian Electricity Act 1910, the Electricity (Supply) Act 1948 and the Electricity Regulatory Commissions Act 1998, governing the power sector in

India. The Act laid down the road map to a liberal framework for operation of power industry, rationalization of electricity tariff and, transparent policies, efficiency in operation, environment friendly policy, creating competition, protecting the interests of consumers and inclusion of all in the ambit of electricity supply. The highlight of the Electricity Act 2003 was provision of national electricity policy, mandatory creation of SERCs, emphasis on rural electrification, open access in transmission and distribution etc. It established Applet tribunal and empowered regulatory commissions to regulate the tariff and issues of license. In nut shell, this Act focused on laws relating to Generation, Transmission, Distribution, Trading and uses of electricity. The Electricity Act 2003 consolidates all the existing laws and introduces provisions with respect to new developments in the sector [15].

The Electricity Act 2003 has laid down the road map for restructuring of Indian Power system [16]. The Electricity Act 2003 has enabled the generators, distribution companies and the consumers to have choice in the matter of supplies of electricity. It specifies the provisions for non-discriminatory use of transmission and distribution system or associated facilities by any licensee, consumer or person engaged in generation.

With the enactment of the Electricity Act 2003, open access was implemented and the market structure in the Indian power sector changed to multi-buyer model of market form the old single buyer monopoly structure. The generators were allowed to sell power to any buyer using the open access provisioned in transmission and users were allowed to choose their power service provider. Since the Electricity Act 2003 has been enacted, the competition among generators and suppliers has been increased in India, which has improved the performance of power sector in India greatly. Currently all those states, who have unbundled the SEBs in their state, have reported improvements in their operational efficiency and have ensured reliable power supply to consumers.

Further the Act was amended on May 28, 2007 and the Electricity Act 2003 was enacted with stronger power and with greater emphasis of assessment, fines as well as legal framework to check the monetary loss due to theft and unauthorized use of electricity.

Reforms in power sector in India are underway to achieve more generation to have more competition among producers and choice for customers. Power sector in India were established as a vertically integrated monopoly organization which had generation, transmission and distribution facilities owned and controlled by the governments. Central government and state governments both were involved in generation and transmission whereas distribution work was mainly done by state governments. Now different generating companies (Independent Power Producers or IPPs) and distribution companies (DISCOs) are also being involved at generation and distribution level.

In the restructured system several new forward looking regulations such as bidding guidelines, open access, flexible tariff regime are considered. Private sector is encouraged to participate in establishment, operation and marketing in emerging power sector due to competitive bidding framework being provided. Foreign investments are coming in a big way in the area of manufacturing of electrical equipment, power plant installation etc. [17]. However, there are problems and technical, operational and commercial issues those demand attention of the policy makers and need to be addressed so that interest of stakeholders can be protected.

The restructured Indian power system may have several markets such as energy market, ancillary service market, transmission service market. It may also have various trading arrangements such as pool model, bilateral dispatch model, pool plus bilateral model and multilateral model [18]. These markets may be classified as perfectly competitive market, oligopoly market or monopoly market, based on the number of suppliers. Indian electricity market may have different pricing schemes such as ex-ante or ex-post, spot price or day-ahead price, system marginal price (SMP), locational marginal price (LMP) or zonal system marginal price (ZSMP) [19]. If conditions allow, the Indian power market may like to introduce retail competition also.

The deregulated Indian power market may have many features and few of them are outlined below:

- Power exchanges and Scheduling coordinator
- Non-discriminatory open access to transmission and distribution network
- Integrated operation of power system like outage planning, relay co-ordination, islanding schemes, etc.
- Central transmission utility and state transmission utilities
- Power pool controller
- Independent System operator
- Load dispatch centres

1.3 Components of Restructured Indian Power System

1.3.1 *Central Electricity Regulatory Commission (CERC)*

Under a new system of mixed entities comprising private and public firms, regulatory bodies at Centre/State level should provide a level playing field to all generating firms, reduce system costs and protect consumer interests. In July 1998, a CERC has been created to set tariffs and regulate inter-state power exchange, licensing, planning and other functions for all central generation and transmission utilities. The role of CERC in the states is largely advisory. It cannot overrule on State Electricity Regulatory Commission (SERC) [18]. All appeals against a SERC decision can only be handled by the State High Court.

- CERC should be empowered to fix the tariff for generation and transmission

- CERC should be authorized to enforce the rules and regulations.

- CERC should be able to finalize the matter regarding inter-state power exchange.

1.3.2 *Independent System Operator (ISO)*

In India at national level an ISO named as Power system Operation Corporation Ltd. (POSCO) – a wholly owned subsidiary of Power Grid Corporation of India Ltd [20]. (PLCIL) has been set up which is the supreme entity in the control of the transmission system. ISO should be disassociated from all market participants, and abstain from any financial interest in the generation and distribution business. However, there is no requirement, in the context of open access to separate transmission ownership and operation. The ISO is not involved in energy markets and its role in generation (or transmission) scheduling will be limited to ensuring that submitted schedules are feasible. The ISO does not perform real-time control of power system facilities, which is being done by Regional Load dispatch Centre (RLDC), State Load dispatch Centre (SLDC) and Area Load dispatch Centre (ALDC) that are hierarchically dependent on it. However, it monitors system operation to ensure adequacy of available reserves, and auxiliary services [21]. It coordinates measures to alleviate transmission congestion and performs contingency analysis to ensure system security against credible contingencies. The different components of Indian power system control structure and their coordination are shown in fig. 1.1.

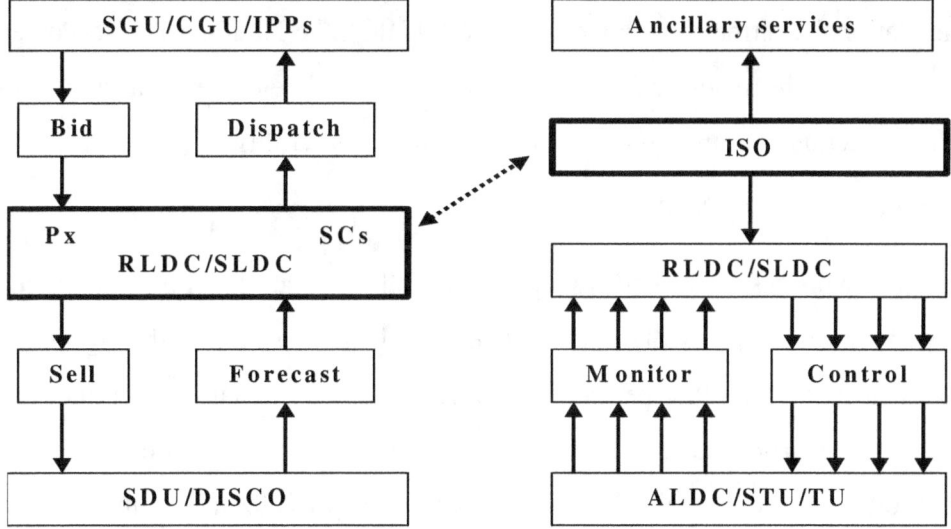

Fig. 1.1: Indian Power System Control Structure

1.3.3 *Transmission Utilities (TU)*

PGCIL started functioning as transmission unit (TU) w.e.f. August, 1991 with the mission of establishment and operation of regional and National Power Grid to facilitate transfer of power within and across the region with reliability, security and economy on sound commercial principals. As on 28th February 2015, PGCIL own and operate more than 1,13,841 Ckt-km network of transmission lines with 2,20,894 MVA transformation capacity and 189 nos. substations that constitutes most of India's interstate and inter-regional electric power transmission system and carries electric power across India. Total Inter-regional power transfer capacity of PGCIL is 45,850 MW. PGCIL transmits mostly at 400 KV, together with 220 KV and some 132 KV ac transmission system criss-crossing the entire length and breadth of the country. HVDC back to back station at Vindhyachal and Chandrapur and a HVDC bi-pole Link (830 km) between Rihand and Dadri are also under operation. PGCIL is maintaining transmission system availability of over 99% and earned the unique distribution of being ranked amongst best transmission utilities in the world. PGCIL also maintains 29,641 kms. of telecom network and it has its point of presence at 370 locations throughout the country with intra-city network of 68 cities across India [22].

1.3.4 *Central Generating Utilities (CGU)*

The National Thermal Power Corporation (NTPC), the National Hydro Power Corporation (NHPC) and the Nuclear Power Corporation (NPC) etc are central generating utilities which generate and supply power in bulk to the State Utilities [23].

1.3.5 *Power Pool Controller (PPC)*

The power pool is controlled by a pool controller at state level and the price of electricity supplied through the pool may be determined on one hourly during peak hours and 3 hourly basis during other period depending on system demand, bid prices and availability of generation and transmission. Whole sellers, retailers, or contestable customers buy electricity from the pool. The power pool thus provides a trading mechanism linking generators, wholesale and retail authorities and customers [24]. A separate unit can be created in the states generation corporation to work as power pool controller. PPC should have following responsibilities:

- Accept bids from market participants and bids can be invited/ accepted from market participants from other state also.

- Calculate the pool price for every hour trading interval for peak load periods and for every 3 or 4 hours for other periods.

- Calculates the amount payable by participants.

- Disclose the market at activities like the current pool price, successful bidders.

- Prepares the economic power dispatch schedules.

1.3.6 *State Electricity Regulatory Commission (SERC)*

State Electricity Regulatory Commission has been up which works as the apex body in the state to finalize the policy matters regarded the Electricity Power Market. In all States, SERC, have been established and they are statutorily responsible for efficient and economic development of the power sector in their respective States/Union Territories. SERCs are authorized to fix the distribution tariff, which could be inclusive of the market forces to be determined at an appropriate time in future through Govt. directions [25].

1.3.7 *State Utilities (SUs)*

The SEBs, created under the ES act, have monopoly right to generate and distribute power in their respective states. The constitution and composition of SEBs, their power, operation, staff, financial accounts, and audit procedures are comprehensively covered under the ES act. The ES act empowers state government to give directions on policy matters, if necessary, to guide the functioning of the boards. The issue of what constitutes policy has been left open to interpretation. In practice, the directions from the State government are not just limited to matters of policy but extend to operations [26].

1.3.8 *Regional Load Dispatch Centre (RLDC)*

A system operator at regional level has been established which comprises several states in its region. Hierarchically it is dependent on NLDC and hence to follow the instructions issued by NLDC. It is responsible to monitor and control the scheduling/ dispatch of power amongst the states in the region. It establishes better coordination in the states for efficient and economic use of Inter-State tie lines. RLDC may direct the SLDC, interstate generating stations to increase or decrease their power generation or drawl in contingency situations such as overloading of lines or transformers, abnormal voltage levels leading to threat security of power system [27]. RLDC makes planning to achieve the best optimization in case of outages of generation and transmission systems. In case of inter-state generating stations projects it may coordinate the bilateral agreements to identify the share of states. RLDC also ensures the frequency linked load dispatch and unscheduled interchanges [28].

1.3.9 *State Load dispatch Centre (SLDC)*

An Independent system operator at state level should be established which would look after the system operation and performance control [29]. This can be performed by the state transmission utilities or a separate wing can be established.

The SISO shall have the following responsibilities:

- Issues dispatch instructions.

- Monitor the economic power dispatch schedules.

- Ensures the reliability of the system, transmission line maintenance and proper pricing.

- Ensures proper system planning.

The SISO should have following objectives:

- Protection of consumers from the exercise of monopoly power;

- Reduction of congestion and reflection of line losses.

- Enhance inter-state power transfer facilities.

- The creation of an even-handed regulating structure for all generators and consumers.

1.3.10 *Power Exchange (PX)*

The PX handles the power pool which provides a common ground to match electric energy supply and demand based on competitive bidding. The time horizon of decision making of the pool power market may range from an hour to a week or longer [30], [31]. The day-ahead market is used in many electricity markets where energy trading is done one day before each operating day. An hour-ahead market is also used to take care of the short fall in energy at shorter time scale.

1.3.11 *Scheduling Coordinator (SC)*

Scheduling coordinators would be involved in aggregating the market participants in the energy trades and these participants may enter a market being operated by SC with its own rules which may be different from pool rule leading to different market strategies. Scheduling coordinators may also directly bid or self-schedule resources as well as handle the settlement process [32].

1.3.12 *State Transmission Utilities (STU)*

State Transmission utilities operate and maintain the whole transmission network spread over the whole state. They keep upgrading and extending the transmission network to provide transmission services with better efficiency and security.

1.3.13 *State Distribution Utilities (SDU)*

The distribution network of the state would work for power distribution in state and would be owned and operated by utility itself. Latter on the distribution network ownership and its operation and control can be separated and retailers may be allowed to use this distribution network.

1.4 Proposed Restructured Power System Models in India

Indian Power Industry has varying degrees of monopoly, competition and choices available. They correspond broadly to real electric power systems in India. Many models are implemented or have been proposed for restructuring of Indian Power system which may vary in their actual arrangements at the time of implementation [33]-[34]. In these models the states would have full operational autonomy. The states will have their power pool and the regional grids would be operated as loose power pools. State Utilities would have the responsibility for follwoings:

(i)   Scheduling/ dispatching their own generation (including the generation of their captive licenses).

(ii) Regulating the demand of their customers

(iii) Scheduling their draw according to their share from the inter-state generating stations.

(iv) Arranging the bilateral inter-changes

(v)   Regulating their net drawl for the regional grids as per central/state guidelines.

`

1.4.1   State Pool as Purchasing Agent Model

The model is shown in fig. 1.2. In this model independent generators (IPPs) are permitted. These may be created from existing utilities or they may be new producers who enter the market when a new plant is needed. All generating stations must sell their power, to a power pool which is turn sells, it to state distribution utilities or distribution companies in the service area. This model allows competition in generation. All power generated by State generation companies (SGU), Central generating units (CGU) and Independent power producers (IPPs) sell their power to a power pool called as purchasing agency. State distribution utility (SDU) and distribution companies (DISCOMs) are only able to purchase from the purchasing agency. They do not have a choice of the purchasing of power.

Power dispatch contracts (Power Purchase Agreements or PPAs) are done through power pool. Normally, these contracts have an availability payment, designed to cover fixed costs, and energy, set to cover the variable costs of generation.

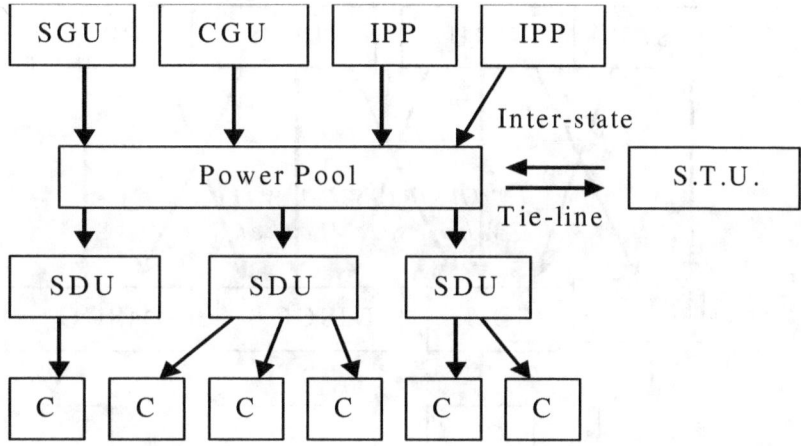

Fig. 1.2: State Pool as Purchasing Agent Model

In this model, sale from purchasing agency to retailers take place at a pre-set tariff price. Efficiency considerations suggest that this tariff should follow the marginal cost of the system while at the same time covering the total costs to the purchasing agency of purchasing power. This tariff should then be modified appropriately from time to time. Retail tariffs, in a competitive retail market, would inevitably tend to follow the cost of purchasing at the purchasing agency, wholesale tariff. This model can accommodate the social obligation policies to be implemented by the government.

In this model construction, which is the most important area to control cost, has been opened to competition. The costs up to the customers can be suitably maintained. Transmission and distribution network can be owned and operated by State and Regional transmission utilities. Inter-state tie line should be sufficient to maintain a loose regional power pool.

1.4.2 State Pool with Wholesale Competition Model

This model provides the choice of supplier for SDU and DISCOs together with competition in generation. SDU and DISCOs can purchase energy for their customers from any competing generator.

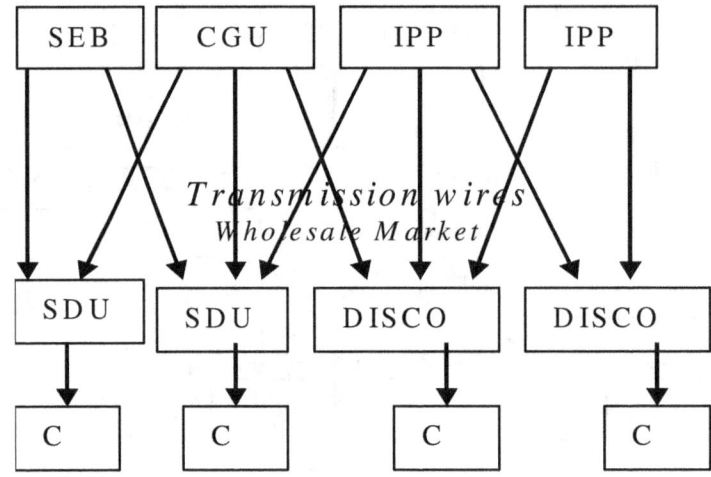

Fig. 1.3: State Pool with wholesale competition Model

These distribution companies maintain a monopoly over energy sales to the final consumers and each of them has a franchise to serve a given set of customers. It requires "open access" to the transmission network, and the development of a spot market. The purchasing agency concept has come to the low-voltage level rather than at the high voltage level but now it is not a single buyer model. Generators may sell directly to any distribution company but open access to low-voltage wires is not permitted.

Since this model permits open access to the transmission wires, it gives the IPPs alternative buyers. It is not therefore necessary for the buyer to take all the market risk, and the form of contract for power can change from the model 1 contract to a contract, which simply hedges price risk. However, customers within a service area still have no choice of supplier. With this model the "obligation to supply" will move to the SDU and DISCOs, which still has a monopoly over the customers. They own and operate the distribution wires.

The transmission network can be owned and maintained by STU, PGCIL and Private Transmission Companies. System operators (SISO, RISO and ISO) should manage the operational control. This model only requires transmission prices for the high voltage wires. These prices must provide the right economic incentives for plant location and dispatch, and sufficient revenue for the transmission owners. In this model the ability to accommodate social policy obligations virtually disappears.

In this model with relatively few customers all of them regulated SDUs and DISCOs, a spot market can be preferable but not essential. It may be possible to be done with some form of regulated open access across a utility's system, with imbalances settled at a terrified rate.

1.4.3   State Pool with Retail Competition Model

In this model all customers have access to competing generators either directly or through their choice of retailer. This would have complete separation of both generation and retailing from the transport business at both transmission and distribution levels. The transmission and distribution wires provide open access. There may be free entry to generation markets and free exit. This means there should be no regulation over "need for new plants" and no requirement to maintain capacity in production when it has passed its economic life. There would also be free entry for retailers. Retailing is a function in this model, which does not require the ownership of the distribution wires although the owner of distribution wires can also compete as a retailer.

This model is not a single buyer model and the power pool in this model are not like purchasing agencies, they are like auctioneers. They never own the power, they do not take the market risk, and they cannot discriminate the price. It should behave like a single transporter, moving power to facilitate bilateral trading. All the trading of power will be done through an integrated network of wires.

The operator of wire should measure and account for the power trades. In this pooling arrangement, there should be provision for bidding into a spot market to facilitate merit order dispatch. The pool will match the supply and demand and determine the spot price for each hour of the day. It collects money from purchasers and distributes it to producers.

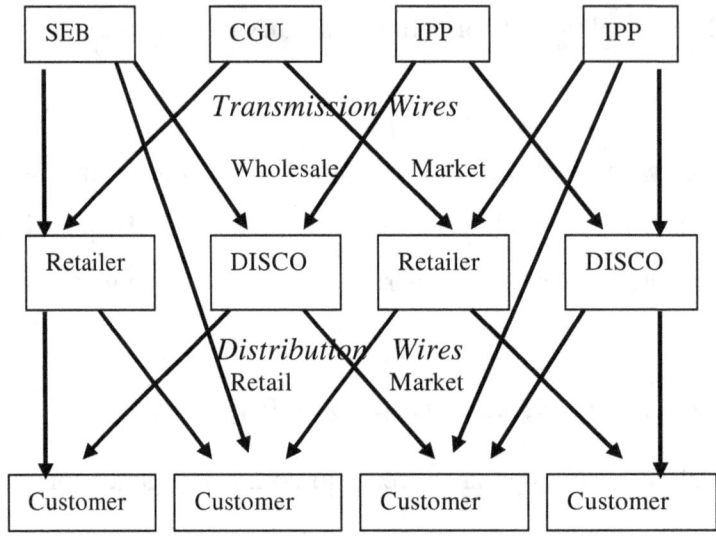

Fig. 1.4: State Pool with Retail Competition

The advantage of these models over monopoly utilities is that competition is introduced in the construction of generating plant. Minimum cost generation can be achieved through competitive bidding for construction and operation of plants, on long-term contracts. Moving from monopoly utilities to this model requires power purchase agreements (PPAs), a legally enforceable contract for the sale of energy from an IPP to a purchasing agency, where dispatch of generating capacity is centrally organized.

However, in this model spot market will become essential, since contractual arrangements between customers and producers are carried out over a network owned by a third party. The network owner must ensure that there are commercial arrangements that allow for the settlement of imbalances between contracted amounts and actual flows.

If different parts of the network are operated separately, inter-area payment schemes will also have to be devised. Metering becomes a major problem in this model. Metering by time of use in no longer merely a useful way of promoting efficient usage but it is a commercial necessity. Each customer needs to be metered on hourly basis, if this is the settlement period. Since the price may change every hour, it is necessary to know how much the customers of each competing retailer used in each settlement period, in order to be able to bill the right customers and to settle accounts properly. If the customers have adequate metering, there will be no problems. But if the numbers of customers are increasing and metering capability for all the customers is not sufficient, it may create logistical problem and provoke disputes.

1.4.4 *Pool with bilateral and multilateral trade model*

In trading with pool model all generating stations must sell the power to power pool and all the distribution utilities/ companies should purchase power only from the pool. Sellers and buyers may enter into bilateral contracts at later stage where the power traded and the prices are at the discretion of the involved parties.

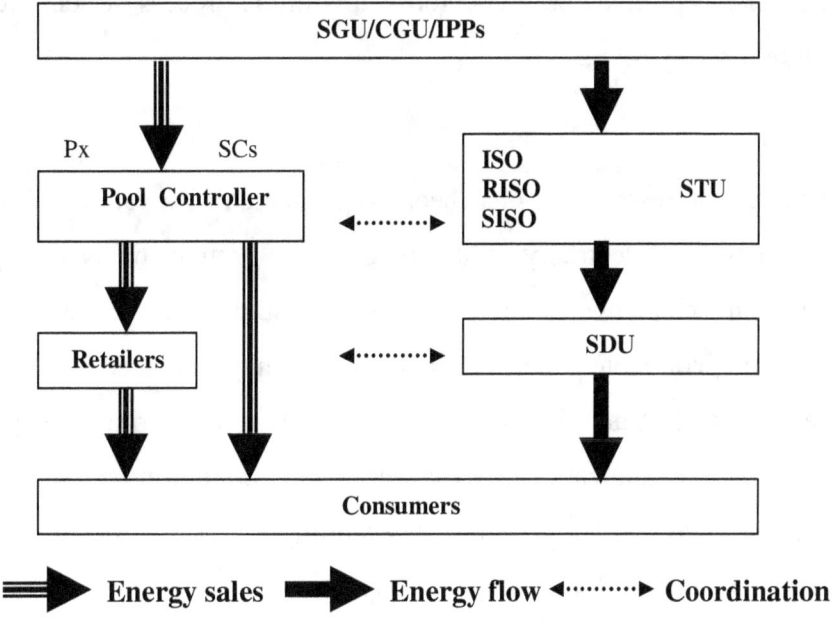

Fig. 1.5: Pool with bilateral and multilateral Trade

The bilateral transactions can further be generalized to achieve multilateral trade where a scheduling coordinator (SC) or power broker can put together a group of energy producers and buyers to form balanced transactions. Bilateral and multilateral trades are allowed with the constraint of static and dynamic security. Bilateral and multilateral transactions can co-exist with a power pool, as shown in fig. 1.5.

1.5 Summary

The Indian electricity Act 2003 aims to create a liberal framework for development of the power industry, promoting competition, protecting interests of consumers and supply of electricity to all areas, rationalization of electricity tariff and ensure transparent policies and promotion of efficiency in Indian Power System. The Act focused on laws relating to Generation, Transmission, Distribution, Trading and uses of electricity. With the enactment of the Electricity Act 2003 and implementation of Open access, the market structure in the Indian Power sector started changing from the old single buyer structure to a multi-buyer market model. In recent days many states, which have unbundled the SEBs, have reported improvements in their operational efficiency and are able to ensure reliable power supply to consumers. The Indian power sector is now in the process of being restructured towards free market based system.

At the same time restructuring has been accompanied by a variety of new problems as the changing nature of electricity utility industry has brought many new practices, technical and operational challenges to power system operation. The success of the new deregulated power system completely depends upon creation of competition in power market. To achieve this there may be different ways of restructuring the power industry but considering the organizational set-up, financial condition, control structure and their coordination many models are in place or proposed for implementing it gradually as the system gets matured in India.

Many new models may be adopted by states based on the policies of state and central government as well as social and market needs of the state for more efficient, secure and reliable power system. A better restructured power system with high degree of competition would provide cheaper and good quality electricity to the consumer and also increase the profit of all market entities.

References

[1] Singh, Anoop (2006), "Power Sector Reform in India: current issues and prospects", Energy Policy, Vol. 34, No. 16, pp. 2480-2490.

[2] http://www.cea.nic.in/

[3] http://planningcommission.nic.in/aboutus/committee/wrkgrp12/wg_power1904.pdf

[4] Central Electricity Regulatory Commission Website (http://www.cercind.org).

[5] www.pfcindia.com/

[6] Rudnick, H. (1996), "Planning in a deregulated environment in developing countries: Bolivia, Chile, and Peru", Power Engineering Review, IEEE, Vol. 16, No. 7,pp. 18-19.

[7] Bacon, R. W. and Besant-Jones, J. (2001), "Global electric power reform, privatization, and liberalization of the electric power industry in developing countries", Annual Review of Energy and the Environment, Vol. 26, No. 1, pp. 331-359.

[8] Porter, M. E. (2011), Competitive advantage of nations: creating and sustaining superior performance, Simon and Schuster, NY

[9] Joskow, P. (2008), "Lessons learned from electricity market liberalization", The Energy Journal, Vol. 29, No. 2, pp. 9-42.

[10] Megginson, W. L. and Netter, J. M. (2001), "From state to market: A survey of empirical studies on privatization", Journal of Economic Literature, April issue, pp. 321-389.

[11] http://powermin.nic.in/

[12] Khaparde, S. A. (2004), "Power sector reforms and restructuring in India", Power Engineering Society General Meeting, IEEE

[13] http://www.cea.nic.in/reports/electricity_act2003.pdf

[14] http://www.mserc.gov.in/acts/no5_electricity_act_2007.pdf

[15] http://www.cea.nic.in/reports/powersystems/nep2012/generation_12.pdf

[16] Singh, R., & Sood, Y. R. (2011), "Current status and analysis of renewable promotional policies in Indian restructured power sector—A review", Renewable and Sustainable Energy Reviews, Vol. 15, No. 1, pp. 657-664.

[17] Bhattacharyya, S. C. (2003), "Review of the Electricity Act 2003 of India", Oil, Gas & Energy Law Journal (OGEL), Vol.1, No. 5, pp. 23-39

[18] http://www.cercind.gov.in/

[19] Aggarwal, S. K., Saini, L. M. and Kumar, A. (2009), "Electricity price forecasting in deregulated markets: A review and evaluation", International Journal of Electrical Power & Energy Systems, Vol. 31, No. 1, pp. 13-22.

[20] http://posoco.in/

[21] A S. A. Khaparde, Power sector reforms and restructuring in India, http://dspace.library.iitb.ac.in/jspui/bitstream/10054/260/3/30010.pdf

[22] http://www.powergridindia.com/

[23] http://powermin.nic.in/

[24] Shym Wadhera and Robin Mazumdar, Evolution of regional power pools in India – current status and trends, Energy Conversion Engineering Conference, IECEC96, Proceedings of the 31st Intersociety (Volume:3), DOI:10.1109/IECEC.1996.553337

[25] http://www.uperc.org/Default2.aspx

[26] http://planningcommission.nic.in/reports/genrep/arep_seb11_12.pdf

[27] http://wbsldc.in/docs/Grid_code_Regulation_2007.pdf

[28] http://www.nrldc.in

[29] http://www.delhisldc.org/

[30] http://www.cercind.gov.in/2014/MMC/AR1314.pdf

[31] http://www.powerexindia.com

[32] https://www.caiso.com/participate/Pages/SchedulingCoordinator/Default.aspx

[33] K.G. Upadhyay, "Design of electricity utility restructuring model for Indian power system", Ph.D. Thesis, U.P. Technical University, Lucknow, India, October 2001.

[34] Foley, A. M., Ó Gallachóir, B. P., Hur, J., Baldick, R. and McKeogh, E. J. (2010), "A strategic review of electricity systems models", Energy, Vol. 35, No. 12, pp. 4522-4530.

CHAPTER 2

Electricity Pricing mechanism and Energy Market Operation in Restructured power System

2.1 Introduction

To achieve the objectives of rapid economic development in India, it is essential to attract good funding in the Energy sector by providing proper return on investment to the different sectors and to ensure availability of electricity to different categories of consumers at affordable rates [1]. It is only possible by having consistency in Regulatory approach and creating proper balance between investments to the Power sector and reasonability of tariff charges. The tariff determination exercise should be based on following criteria mentioned below:

(a) The tariff should genuinely reflects the economic costs and promote efficient usage of electricity

(b) It should improves the financial status of Power utilities, for their future expansion;

(c) It should lead to a suitable mechanism of the electricity tariff adjustment, corresponding with fuel prices.

(d) It should safeguard the interest of all consumers of all categories.

Along with best tariff structure an efficient operating mechanism and structure is must to achieve the objectives of power system restructuring in India. Following are some important requirements of the competitive energy market in India.

- Development of consistent regulatory system along with suitable pricing mechanism

- Creation of sufficient competition in generation

- Investment to create efficient and secure transmission and distribution network

- Open access of Transmission and distribution to all

- Development of suitable structure for day-ahead or real-time market and its efficient management.

- Modelling of energy market with appropriate trading arrangements.

- Development of accurate and fast converging software tools for forecasting the market parameters such as load, price, ancillary services etc.

Rest of the chapter discusses various pricing mechanisms available worldwide including India and constituents of the restructured power system which makes its operation more efficient.

2.2 Present tariff determination system in India

As per section 79 of the Electricity Act'2003, Central Electricity Regulatory Commission (CERC) has been authorized for Tariff determination of Generation Utilities [2]. CERC is also authorized to determine the Tariff for Inter-state transmission of electricity. As per Section 86 of the Electricity Act'2003, State Electricity Regulatory Commission (SERC) are mandated to determine the tariff for Generation, Transmission and Supply of electricity within the all states. As per Section 61 of the Electricity Act'2003, the respective Commission has to stipulate the terms & conditions for determination of Tariff.

2.2.1 *Regulatory Norms for Computation of Tariff for Thermal Power Plant*

The regulatory norms for the computation of tariff for thermal power station are discussed below.

2.2.1.1 Capacity Charges & Energy Charges

For thermal power generating stations (coal, lignite and gas based) the CERC has adopting a two-part tariff [3]:

(a) Components of Capacity Charges/Annual Fixed Charge (AFC)

The Fixed component of the tariff is dependent on the capital cost of the project. The fixed component of the tariff as shown in table 2.1 below ensures that the Power producer is able to recover the fixed expenses and earn a return on investment, irrespective of the actual generation.

Table 2.1: Components of Annual Fixed Costs

Components of AFC	FY 2009-14
Return on Equity	15.50%
Interest on Loan Capital	As per actual
Depreciation	5.28%
Interest on Working Capital	Based on Normative Parameters
Operation & Maintenance Cost	Based on Normative Parameters
Cost of Secondary Oil	Based on Normative Parameters
Special allowance in lieu of R&M	Based on Plant life

(i) Return on Equity

SERC has specified a Pre-Tax Return on equity (RoE) of 15.5% for the Tariff period FY 2009-14. Further, it has allowed an additional RoE of 0.5% for projects commissioned after April 2009 within specific timelines.

(ii) Interest on Loan Capital

The CERC has specified a debt-equity ratio of 70:30 as the funding mix for the capital cost of a project. The interest rate as per actual on these loan funds is recoverable as part of the tariff. The Tariff Regulations allows retention of $1/3^{rd}$ of the benefits, if any, arising out of re-financing of loans; earlier such benefits were required to be passed on entirely to the beneficiaries.

(iii) Depreciation

Based on a 25-year project life and 90% of the capital cost CERC has fixed the depreciation rate to 5.28% for most components of the project.

(iv) Interest on Working Capital

CERC has specified the norms for Coal stock, Secondary Fuel Oil Stock Maintenance spares, Sales receivables and O&M expenses as shown in table 2.2 below.

Table 2.2: Working Capital Components – Thermal Plant Tariff

Components	FY 2009-14
Coal Stock	$1^{1/2}$ Months for Pit Head 2 Months for Non-Pit Head
Secondary Fuel Oil Stock	2 Months
Maintenance Spares	20% of O&M Cost- Coal Based 30% of O&M Costs-Gas Based
Sales Receivables	2 Months
O&M expenses	1 Month

(v) Operations & maintenance costs

The CERC has specified O&M costs for thermal power stations on the normative parameters (Rs. Lakh/MW), depending on the class of the machine installed by the power station. The allowed normative O&M expenses are presented in table 2.3 below.

Table 2.3: Normative O&M expenses – Thermal plant tariff

Rs Lakh/MW	200/210/250 MW	300/330/35 MW	500 MW	600 MW And Above
2009	18.20	16.00	13.00	11.70
2010-11	19.24	16.92	13.75	12.37
2011-12	20.34	17.88	14.53	13.08
2012-13	21.51	18.91	15.36	13.82
2013-14	22.74	19.99	16.24	14.62

(vi) Cost of secondary fuel oil & limestone

As per tariff regulations for the period FY 2009-14, the CERC has included the cost of secondary fuel oil (SFO) as part of AFC.

(vii)   Special allowance in-lieu of R&M

CERC has granted an allowance to the coal/lignite based thermal power project @ Rs 5 Lakh/MW/Year from 2009-10 and escalated @ 5.72% p.a.

(b) Energy Charges (for recovery of Primary fuel costs)

Energy charges for thermal power stations are linked to the normative operational parameters as specified by the Regulator. The approved aggregate gain or loss the Distribution licensee on account of Uncontrollable factors in passed through, as an adjustment in the tariff of the Distribution licensee.

2.3 Methodologies for Electricity price determination in restructured power system

The suitable pricing schemes will enable the investors to take decision regarding the timing, magnitude and type of fuel for new investment and can have maximum incentive to produce their output at the cost and can earn higher profits by cost reducing innovations. In the restructured electricity supply industry there are different market rules and market structures in different countries [4]. These market rules and structure of the various industries may enable to determine whether these markets set economically efficient prices. Many pricing methodologies are available and same are discussed below.

2.3.1 *Marginal Cost Methodologies*

Economists argue that in a regulated environment, marginal cost based pricing of goods and services provide an efficient economic solution to developing a tariff structure. The short-run costs of wheeling are the marginal (incremental) costs of the last MWh of energy wheeled. Since wheeling is physically indistinguishable from a simultaneous purchase-sale by the wheeling utility, the short-run marginal wheeling costs can be computed from the marginal costs (spot prices) of electricity at the buses where it enters and leaves the wheeling utility, and ideal wheeling rate is the marginal cast of wheeling. The allocation of marginal cost is done through the usage calculation. On an annual basis, the marginal cost of a transmission service transaction can be defined as [5]:

$$MCC = \sum_{f \in F_N} \frac{\left|\Delta MW_{f,l}\right| \times MC_f}{\sum_{s \in S_N} \left|\Delta MW_{f,s}\right|} \tag{2.1}$$

where, MCC is the annual Marginal Capacity Cost,

$\Delta MW_{f,l}$ is the change in megawatt flow due to the contracted transmission service on new facility f

$\Delta MW_{f,s}$ is the change in megawatt flow due to transmission service on new facility f for all marginal sales s,

MC_f is the cost of new facility f which is the sum of depreciation on facility f, marginal cost of capital, marginal taxes and marginal expenses for any year of the transaction,

F_N, S_M are the sets of all new facilities and marginal sales, respectively.

2.3.1.1 Marginal Operating Cost

The marginal operating cost is the incremental cost of the last MWh of energy in optimal operating conditions. The marginal operating costs are based on the cost of fuel, transmission losses and the operational costs for dealing with generation and line capacity limits. The OPF is a technique for minimizing the total cost of operating the spatially distributed system subject to the satisfaction of all load flow equations and system operational constraints. The solution provides the optimal operating state and the dual variables give marginal operating costs and shadow prices corresponding to binding constraints. The marginal costs of real and reactive power at bus i, MC_{pi} and MC_{qi} respectively, are [6]:

$$MC_{pi} = \frac{\partial}{\partial P_i} \tag{2.2}$$

(total generating and wheeling costs subject to the operational constraints)

$$MC_{Qi} = \frac{\partial}{\partial Q_i} \tag{2.3}$$

(total generating and wheeling costs subject to the operational constraints)

These marginal costs, MC_{pi} and MC_{qi} are equal to the Lagrangian multipliers of the corresponding Kirchhoff equations when the OPF is solved as a non-linear programming problem.

2.3.1.2 Marginal Cost Based Wheeling Rates

Marginal cost based wheeling revenue is the cost of wheeling the scheduled real and reactive power. If the seller is connected at S and the buyer at B, the wheeling rate will be as given below.

$$\omega_p = MC_{PB} - MC_{PS} \tag{2.4}$$

$$\omega_Q = MC_{QB} - MCQS \tag{2.5}$$

$$W_P = P_B\, \omega_p \tag{2.6}$$

$$W_Q = Q_B\, \omega_Q \tag{2.7}$$

$$W = W_P + WQ \tag{2.8}$$

where ω_p and ω_Q are wheeling prices for real and reactive power, respectively, while Wp and W_Q are the total wheeling charges for real and reactive power, W is the total wheeling charge. MC_{PB} and MC_{QB} are marginal costs of real and reactive power, respectively, at the buyer bus while MC_{ps} and MC_{QS} are the values at the seller bus. P_B and Q_B are real and reactive wheeling power at B.

As power system margins are reduced and the operations are much closer to their technical limits, marginal costs of reactive power and the effect of reactive power flow on real power marginal costs should be included in the consideration of wheeling rates while the establishment of a wheeling rate needs to allow for modifications in reactive resources.

2.3.1.3 Short Run Marginal Cost (SRMC)

Conceptually, the short run marginal cost p, of active power in a bus i in a system can be calculated as:

$$\rho = \gamma + \gamma \frac{\partial L}{\partial d_i} - \sum \mu \frac{\partial Z}{\partial d_i} \tag{2.9}$$

where d_i is the demand in i, γ is the Lagrange multiplier associated to the power balance equation, μ is the vector of Lagrange multipliers associated to network constraints, L are the network losses and Z are the network restrictions: line thermal transfer limits, security restrictions, maintenance, etc. Therefore, the marginal cost in any bus is made up of two components.

γ that represents marginal generating cost, called "system lambda", and $\gamma \frac{\partial L}{\partial d_k} - \sum \mu \frac{\partial Z}{\partial d_k}$ called "lambda differential", which varies with the bus and is dependent on two subcomponents, one associated to marginal cost of losses and the second to network constraints. This second term can be zero if no constraints are present but it becomes very important if limits are reached.

In other methods [7], the marginal operating cost of the power system due to a transmission transaction is calculated first. The marginal operating cost is then multiplied by the magnitude of the transacted power to yield the SRMC for the transmission transaction.

$$SRMC_t = \sum_{i \in B_t} BMC_i \times P_{i,t} \qquad (2.10)$$

Where, BMC_i is the bus i marginal cost, $P_{i,t}$ the injected power at bus i due to transaction t and B_t the set of transaction buses involved in the transaction t.

SRMC prices may not closely follow a transmission transaction actual operating cost if the magnitude of the transacted power is large compared to the magnitude of active load in the transmission system.

2.3.1.4 Long Run Marginal Cost (LRMC)

The LRMC usually refers to the costs of incremental production over a long period, including both variable and fixed costs and can be defined as the marginal cost of supplying an additional unit of energy when the installed capacity of the system is allowed to increase optimally in response to the marginal increase in demand. As such it incorporates both capital and operating cost. For the system as a whole, a single value of LRMC can be developed for a time period which reflects the cost of system expansion.

Four major weaknesses in the present-day LRMC calculation are [8]:

- Computations are cumbersome since they involve multiple solutions of usually sizeable mathematical programming models.

- Demand models used are often oversimplified.

- The marginal cost determination depends on the huge uncertainties in the future values of important inputs such as load growth, fuel costs, cost of capital, construction times, etc, are ignored.

- Most present-day LRMC approaches deal only with future capital and operating costs. They do not include the impact of a change in demand d(t) on the operating costs at hour t.

2.3.2 Embedded Cost Methodologies

Embedded cost can be defined as revenue requirements needed to pay for all existing facilities plus any new facilities added to the power system during the life of the contract for transmission service. The allocation of the embedded cost can be done through the usage calculation. On an annual basis, the embedded cost of transmission service transaction can be defined as [9]:

$$ ECC = \sum_{f \in S} \frac{\left| \Delta MW_{f,l} \right| \times EC_f}{\sum_{s \in S} \left| \Delta MW_{f,s} \right|} \tag{2.11} $$

where,

ECC is the annual Embedded Capacity Cost,

$\Delta MW_{f,l}$ is the change in megawatt flow due to the contracted transmission service l on facility f,

$\Delta MW_{f,s}$ is the change in megawatt flow due to transmission transaction s on facility f,

EC_f is the annual embedded cost of facility f which is the sum of depreciation, embedded cost of capital, taxes and expenses. S, F are the sets of all sales S and facilities F in a given year.

Various embedded cost of wheeling methods are presented below.

2.3.2.1 Rolled-In-Embedded Method

The Rolled-in-embedded method assumes that the entire transmission system is used in wheeling, irrespective of the actual transmission facilities that carry the wheeled power [38]. The magnitude of the transacted power for a particular transmission transaction is usually measured at the time of system peak load condition:

$$R_t = TC \quad \text{x} \quad \frac{P_t}{P_{peak}} \tag{2.12}$$

where R_t is the transmission pricing for transaction t, TC is the total transmission charges and P_t and P_{peak} are transaction t load and the entire system load at the time of system peak load condition, respectively.

Since this method ignores the actual system operation, it is likely to send incorrect economic signal to transmission customers. As a result, a wheeling transaction that may require extensive and costly system upgrades can still take place because the wheeling customer will be responsible for only the fraction of the transmission related cost caused by the transaction.

2.3.2.2 Contract Path Method

The contract Path method is based upon the assumption that the wheeling is confined to flow along an electrically continuous path through the wheeling company's transmission system. The embedded capital costs, correspondingly, are limited by those facilities which lie along this assumed path. A portion or all charges associated with transmission facilities in the contract path are then allocated to the wheeling customer. If new transmission facilities are built for the wheeling transaction, they are usually included in the contract path. Considering some simplifying assumptions this method can be interpreted as a solution to the optimal transmission-planning problem from the static point of view.

2.3.2.3 Boundary Flow Method

The Boundary Flow or Power Allocation Method (PAM) equates the impact of a sale on transmission system to the gross change in real power outflow from the system caused by the sale. A rate is calculated by normalizing by the change in real power outflow by the magnitude of the transaction and then multiplying this ratio by the system transmission cost. The multiplying ratio is called the PAM factor and is usually capped at unity.

$$PAM\ Cost = \frac{\sum_{t \in T}(Flow_{final,t} - Flow_{initial,t})}{Magnitude\ of\ Transaction} \times ARR \tag{2.13}$$

Where,

ARR is the total transmission annual revenue requirement for the system, and T is the set of all ties. It is worth noting that the sum of PAM factors is unlikely equal to unity. Hence it is not advisable to use PAM for other methods of cost allocation.

2.3.2.4 MW-mile Method

The MW-mile method measures the amount of transmission capacity used by summing the products of each facility length and the change in flow quantity on the transmission facility caused by the sale. Many alternatives have been used to measure the facility's contribution to the capacity of the system which includes temperature based ratings, Surge impedance Loading, actual power flow etc.

The MW-mile method first calculates the flow on each circuit caused by the generation load pattern of each agent based on a power flow model. Costs are then allocated in proportion to the ratio of power flow and circuit capacity:

$$R(u) = \sum_{all\ k} C_k \frac{|f_k(u)|}{\overline{f}_k} \tag{2.14}$$

where,

R(u) is the allocated cost to agent u,

C_k is the cost of circuit k,

$F_{k(u)}$ is the k-circuit flow caused by agent u,

f_k the k-circuit capacity

$$\text{Total cost} = \sum_{all\ k} C_k$$

As the total circuit power is usually smaller than the circuit capacities, this allocation rule does not recover all embedded costs. In terms of transmission expansion interpretation, this means that the MW-mile scheme is only charge for a "base-case" network, but not for "transmission reserve", given by the difference between circuit capacity and actual flow.

This methodology neglects the actual system operation but overcome some limitation of the rolled-in methods.

2.3.3 Incremental Cost Methodologies

Incremental cost can be defined as the revenue requirements needed to pay for any new facilities that are specifically attributed to transmission service customers. These facilities must be identified for all year across the line of contract for transmission service. This includes revenue requirements in years beyond the life of the contract. If a facility is needed by more than one transmission service customer; then the cost of the facility can be allocated to the incremental customers by the usage method [10].

$$ICC = \sum_{y \in Y} \sum_{f \in F_I} \frac{\left|\Delta MW_{f,l,y}\right| \times IC_{f,y}}{\sum_{s \in S} \left|\Delta MW_{f,s,y}\right|} \times PWF_f \qquad (2.15)$$

where, ICC is the total incremental capacity cost across the life of the contract.

$\Delta MW_{f,l,y}$ is the change in megawatt flow due to the contracted transmission service i on incremental facility f for the year y,

$\Delta MW_{f,s,y}$ is the change in megawatt flow due to all transmission service on facility f for all incremental customers s in year y that required this incremental facility.

$IC_{f,y}$ is the incremental facility f in year y which is the sum of depreciation of facility f, incremental cost of capital, incremental taxes and incremental expenses.

F_I, S, Y are the sets of incremental facilities, incremental customer sales, and service life years of each incremental facility, respectively.

PWF_y is the appropriate present worth factor.

In the following section, we will discuss the short-run incremental cost (SRIC) and long-run incremental cost (LRIC) methods.

2.3.3.1 Short-Run Incremental Cost (SRIC) Method

This pricing method contains evaluating and assigning the operation costs associated with a new transmission transaction. The transmission transaction operating costs can be estimated using optimal power flow (OPF) that accounts for all operation constraints including transmission system static or dynamic security constraints and generation scheduling constraints. Since revenues collected through this pricing method only compensate for the operating cost incurred by a transaction, SRIC pricing could discourage host utilities from expanding their transmission system.

2.3.3.2 Long-Run Incremental cost (LRIC) Method

This pricing method contains evaluating all long-run operating and reinforcement costs necessary to accommodate a transmission transaction and assigning such costs to the transaction. The operating cost component may be evaluated based on the same principles described in SRIC.

2.4 Availability based Tariff (ABT)

Availability Based Tariff (ABT) has been adopted in India through ABT Order dated January 2000 of CERC. It is a frequency based pricing scheme for electricity supply and it is related with the tariff structure for bulk power. Its aim is to incentives and disincentives the power generators and consumers so that they use power with more responsibility and accountability.

Under ABT mechanism power is charged and regulated to achieve network stability in electricity market. Through this scheme the CERC looks forward to improve the energy sector in terms of power quality, improve high frequency deviations causing damage and disruption to large scale industrial consumers and frequent grid disturbances resulting in power outages, tripping of generators resulting in disintegration of grid. Initially ABT applied to only generating stations under central govt. having more than one state or SEB as its beneficiary but now it has been expanded to cover the intrastate systems also.

The important features of Availability based tariff (ABT) can be summarized as:

- Enhanced grid discipline

- Economically viability of operations

- Competitive pricing

- Technical and commercial efficiency

- Merit Order Dispatch/ Economic Dispatch system

- Less grid disturbance

- Curtailment of market power and gaming

- Smart meters, SCADA based real time data acquisition and control

- Industry standard communication mechanisms and open protocols

- Software applications for computing, implementation of regulatory requirements and its modifications as and when required

- Online interfaces to stakeholders for effectiveness of electricity market

- Better control of production cost and greater flexibility in operations by power producers

Under ABT mechanism, scheduling is done each day starting from 00.00 hours by dividing the 24 hours into 96 time blocks of 15 minutes each. Every participating generator declares its generation capacity in advance in terms of MWh delivery ex-bus for each 15 minute time block of the next day. In case of hydro stations, the total ex-bus MWh which can be delivered on that day will also is published in advance. It is important that at the time of declaration of power to be delivered, the generator ensures that the capability during peak hours is same or more compared to other hours. While declaring the capability, the scheduling should follow the laid down operating procedure strictly.

Based on these declarations, the Regional Load Dispatch Centre (RLDC) would decide and communicate the shares of the available capability to various beneficiaries. After getting the requisition for power based on the generation schedules form beneficiaries, the RLDC would prepare the generation and drawl schedules for each time block considering the technical limitations and transmission constraints. The schedule of actual generation would be quantified on ex-bus basis and for beneficiaries scheduled drawls would be quantified at the receiving points.

The transmission losses are apportioned in proportion to the drawls of beneficiaries while calculating the drawl schedule. The schedules may be revised by RLDC in case of any forced outage or transmission bottleneck. The revised schedule would become effective from the fourth time block, starting from the time block in which the revision was suggested by the generator. Generators and the beneficiaries are also permitted to revise their schedules during the day, but such revisions would be effective only from the 6th time block.

2.5 Transmission open access

Transmission open access (TOA) aims to achieve competition in electric power supply by breaking the monopolistic regulatory structure. It introduces new regulations, rights and obligations, operational procedures and economic conditions enabling participants to use the transmission network for transferring power which may belong to other participants.

TOA makes wheeling of electrical energy as one of the more prevalent unbundled service of electricity market which involves at least three parties (i) seller, (ii) buyer and (iii) wheeling utilities to transmit the power from the seller to the buyer. Transmission service pricing are decided to achieve economic, efficiency, revenue sufficiency and efficient regulation. The costs of transmission transactions have four major components.

- *Cost of production* arising due to generation re-dispatch and rescheduling as a result of transmission transactions.

- *Cost of opportunity* due to operating constraints that are caused during transmission transaction.

- *Reinforcement cost* to have new transmission facilities to accommodate the transmission transaction.

- *Allocation cost* of existing transmission facilities used by the transmission transaction.

TOA implementation may be different in different electrify markets depending on electric system structure and transaction characteristics.

2.5.1 Electric System Structures

Transmission open access may be considered within a single independently dispatched entity, ranging from a single vertically integrated utility to a tight power pool or at the multiple independently dispatched entity level. Transmission may be either vertically integrated or unbundled from generation and distribution. An unbundled transmission company can be more easily regulated to provide network service on a non-discriminatory basis to all its users, while a conglomerate of vertically integrated utilities plus independent power producers (IPPs) and distribution companies will tend to adopt wheeling charges that are based on the total incurred costs by each utility.

The two major extreme regulatory options are the cost-of-service approach and the fully competitive market. Power pools are examples of this non-competitive approach. Some mandatory transmission access amounts to a breach in the traditional approach, resulting in a hybrid scheme where a limited amount of competition coexists with the original regulation. On the other hand, in a fully competitive market, transmission access is intrinsic to the market mechanism.

2.5.2 Transaction Characteristics

Transmission access is typically granted to some, but not all participants in the electricity business. It may be restricted to the operators of the independently dispatched entities, which then may exchange power only among themselves. The right may be granted to all of the electric utilities within an independent dispatched entity. Finally, access may be opened to independent generators and or all customers. In the last two cases, the right may apply only to transactions among independently dispatched entities, or to both.

Transmission access rights may encompass either short-term and/or long-term transactions. Short-term transactions are physical and negotiated and contracted in the time frame of real-time economic dispatch. This is typical of operators of the energy management control centres of independently dispatched entities. Long term transactions may be contracted well in advance and may have duration ranging from weeks to several years.

2.5.3 Economic Issues

The generation business has high degree of freedom in the selection of the energy source and its features are absence of energy storage technologies, long project development time, less significant economies of scale. The resources may be capital intensive with low operation cost or low capital with high operation cost. Transmission business is characterized as a natural monopoly. To operate and expand the business efficiently certain regulation is required to provide adequate returns as well as economic incentives. The business must provide the transmission service based on standards of quality, reliability, and security. Historic acquisition cost, recovering past investments or paying for future ones may be the criteria for costing the assets that provide the transmission services. The distribution business tends to develop through geographic monopolies, although there are no clear economics of scale.

The transmission business is the area where important economics of scale are present, where competition may not be feasible, and natural monopolies develop. Nevertheless, it is the business that benefits the generation and distribution business and also consumer in the end through interconnections which allow generators to compete.

2.5.4 Operational Issues

Many different TOA structures have been developed world-wide, based on reasons such as historical, political, or economical. Regardless of the form TOA takes in different parts of the globe, power system grid control is required to turn TOA policies into practice. TOA policy may affect the optimization and control algorithms, the security functions, and the communication facilities. Depending on a nation's electrical interconnection size and complexity, grid control may be the designated responsibility of a single business entity (e.g., utilities and power pools). This is based on the premise that TOA can be achieved within the framework of "control area", although their current composition (geographic and economic) may need to change. The focus of TOA on power system operations can be viewed according to time frames, before-the-fact, real time, and after-the-fact. These aspects of operations are reviewed from the perspective of traditional practices and in the light of new requirements needed to support TOA.

2.6 Trading arrangements in Electricity Market

The electricity market mechanisms are designed to give commercial freedom to the participants to decide their own operating mechanism and improve the market transparency for achieving higher efficiency in the day-ahead and real-time electricity markets [11]. Its objective is providing equal access to participants of the market and provides equal treatment to all participants in accessing the financial tools to hedge the risks associated with real-time energy market pricing. It provides the mechanisms to manage transmission congestion and offer incentive to participants for building additional generation capacities to cater to market needs.

2.6.1 Day-Ahead and Real-Time Markets

Day-ahead and real-time markets are operated and administered by power pool to facilitate sale and purchase of electricity. It receives and processes the supply and demand bids and manages the state/ regional level market processes, publishes day-ahead schedules and energy clearing prices. It determines and publishes the real-time prices and load zone for each trading interval and identifies load zones and settlement nodes (a point on the grid where bills and credits for participants are produced for the energy received or supplied and nodal prices are calculated) for generators for the purposes of price determination. Pool also lays down conditions for the market suspension and disseminates the necessary information to enable the market to operate efficiently. It also compiles and publish trading statistics of day-ahead and real-time markets.

2.6.1.1 Day-ahead Market

The Power Pool should voluntarily operate a day-ahead market for trading energy on behalf of market participants. The power pool prepares a day-ahead supply offer curve by sorting all of the supply offers in merit order from the lowest price/quantity pair to the higher price/quantity pair. The power pool also prepares a day-ahead demand bid curve by sorting all of the demand bids in merit order from the highest bid price/quantity pair to the lowest bid price/quantity pair. The power pool then compares the supply offer curve and demand bid curve to determine the point of intersection which represents the market clearing point representing the total amount of energy purchased and sold in the day-ahead market. The day-ahead market energy clearing price, for each trading interval of the day should be equal to the price at which the day-ahead market got cleared.

The day-ahead supply and settlement obligations for energy of a participant are estimated based on the amounts cleared in the day-ahead market. Each participant with a supply offer price which is less than or equal to the energy clearing price has a supply obligation equal to the amount of MWs submitted as part of the supply offer that are priced at or below the energy clearing price. Each participant with a demand bid price which is greater than or equal to the energy clearing price has a settlement obligation equal to the amount of MWs submitted as part of the demand bid that are priced at or above the energy clearing price in day-ahead market.

2.6.1.2 Real-time market

In the real-time market there are certain guide lines and timings for submission of required information by participants to the power pool for the development of the hour-ahead schedule and conditions forecast. All the suppliers those are slow start should indicate their self-commit/self-decommit schedules as part the supply offer. The independent system provider (ISO) should accept these schedules to the extent that their schedules do not cause any reliability problem. The pool should develop the hour-ahead schedule based on an optimization process which should be capable to minimize the total cost of energy supply. This is necessary to satisfy accepted demand bids and provide the suitable quantity of contingency reserve in the real-time market. The following process should be followed by the power pool and participants with regard to overall market operations.

- Prior to the trading timeline, participants would submit supply offers and demand bids to the pool using market information system of the pool.
- The pool would evaluate the supply offers and demand bids received and develop the hour-ahead Schedule using a software application immediately after end of dead line time.
- The pool prepares the conditions forecast based on the information submitted prior to the timeline using some forecasting software tool.
- The pool also prepares the provision of AGC and contingency reserve within a trading interval.

The pool should establish a process for validation of supply offers and demand bids prior to their use in the schedule for the day-ahead and real-time markets. The pool should promptly notify the participant if a supply offer or demand bid has been validated or rejected with reason. It should also maintain a log of all supply offers and demand bids that are submitted and with results of validity checks. A participant may submit a new corrected supply offer and demand bid prior to the daily trading deadline. The pool would consider the last valid supply offer or demand bid submitted by a participant as final. Supply offers or demand Bids would remain in effect until withdrawn by the participant.

Power pool determines the hour-ahead schedule for each trading interval on the basis of supply offers and demand bids submitted and publish an hour-ahead schedule covering each trading interval of the period. The hour-ahead schedule includes following information:

- forecasts of the most probable peak system load, required ancillary services and probable Energy consumption for each load zone and for the total system

- aggregate generator availability for each load zone

- forecasts of nodal prices

- Expected energy surplus and deficit for each load zone

- projected shortage of ancillary service requirements

- forecasted timing of when expected conditions may be inadequate plus trading interval for which capacity deficiency conditions are forecasted

The power pool documents the operation of the hour ahead-scheduling process and all such documentation are made available to participants through website of power pool. Each Participant must fulfil its commitment as required under the hour-ahead schedule and is responsible for changing commitment in the real-time Market.

2.6.2 Network Model

The power pool maintains a network model with all details for software based settlement of locational prices to operate the real-time market accurately. It identifies in consultation with market participants set of settlement nodes corresponding to a physical system bus on the grid. If needed, several electrically equivalent system buses may be combined into one settlement node. A settlement node is created for each generator connected to the grid and at all physical locations at which a dispatch able load is connected to the Grid.

At the start a single load zone comprising of the power pool control area boundary is defined by the pool for settling all the energy obligations in the real-time market. Later if participants allow and congestion of grid requires, additional load zones may be created. Additional load zones should be created when supply nodes or demand nodes having large differences in nodal Prices are located in separate load zones. The transfer limits boundaries between load zones are clearly defined and transfer flows across load zones are measured at the boundary. The boundaries of the load zones must be reviewed by power pool if the criteria set are not achievable and if necessary action should be taken to alter the boundaries or formation of new load zone using the similar criteria.

To manage the day to day operation of the market, power pool must execute and publish a forecast about system conditions, determine and schedule ancillary services to eliminate the capacity deficiency and maintain security and reliability of power system. The power pool must provide services for financial settlements including billing and clearance in day-ahead and real-time markets.

2.7 Summary

Restructuring of power system with consistent regulatory system along with suitable pricing mechanism is important to achieve the objectives of rapid economic development in India. This will ensure proper operation of energy market in India with high degree of efficiency. Power system restructuring would turn to be successful only if there is sufficient competition in generation, efficiency in transmission and distribution with security of operations as well as benefit to end users. Restructuring should improve the financial status of power utilities so that they can invest for up gradation and expansion and at the same time safeguard the interest of all consumers of all categories. Open access of transmission and distribution is a must for success of restructuring of Indian power sector. Creation of suitable structure for day-ahead or real-time market and its efficient operating mechanism is the need of the hour. Realistic modelling of energy market need to be built up and new trading arrangements are to be put in place to meet the challenges of Indian energy market. Market operator should develop accurate and fast converging software tools for forecasting the market parameters such as load, price, ancillary services etc. to support the operations of the electricity market.

References

[1] http://cercind.gov.in/2009/February09/SOR-regulations-on-T&C-of-tariff-05022009.pdf

[2] http://planningcommission.gov.in/aboutus/committee/wrkgrp12/wg_power1904.pdf

[3] http://npti.in/Download/Distribution/Tariff%20Determination%20-%20Write%20up.pdf

[4] Sergey Voronin, "PRICE SPIKE FORECASTING IN A COMPETITIVE DAY-AHEAD ENERGY MARKET"
http://www.doria.fi/bitstream/handle/10024/93793/isbn9789522654625.pdf?sequence=2

[5] Khaparde, S. A. (2004), "Power sector reforms and restructuring in India", Power Engineering Society General Meeting, IEEE

[6] Mahbube Zeraatzade, "Transmission Congestion Management by Optimal Placement of FACTS Devices", http://bura.brunel.ac.uk/bitstream/2438/4710/1/FulltextThesis.pdf

[7] S.S.Sankeshwari, Rajashekher Koyyeda, Dr.R.H Chile, "A Unified Approach to Transmission Network Cost Allocation", International Journal of Engineering Research and Applications (IJERA) ISSN: 2248-9622, Vol. 2, Issue 3, May-Jun 2012, pp.2676-2684

[8] http://www.eppo.go.th/power/MarketRules-Draft10-1-19-01.doc

[9] Bacon, R. W. and Besant-Jones, J. (2001), "Global electric power reform, privatization, and liberalization of the electric power industry in developing countries", Annual Review of Energy and the Environment, Vol. 26, No. 1, pp. 331-359.

[10]   Singh, Anoop (2006), "Power Sector Reform in India: current issues and prospects", Energy Policy, Vol. 34, No. 16, pp. 2480-2490.

CHAPTER 3

Introduction to Forecasting in Electricity Market

3.1   Introduction

In the vertically integrated electricity supply industry, load forecasting has been a very important activity in the power system planning and operation but due to introduction of competition in the electricity sector, the forecasting of several variables such as electric load, electric price, spinning reserves, bid prices, etc. is difficult task. The forecasting of these variables is challenging not only due to several market players but also due to imperfect competition in the electricity markets such as energy market, ancillary service market, transmission market, etc. Moreover, various trading arrangements [1] such as pool model, bilateral dispatch model, pool plus bilateral model and multilateral model have different degree of uncertainties. Based on the number of suppliers, market may be classified as perfectly competitive market, oligopoly market or monopoly market.

Different pricing schemes such as ex-ante or ex-post, spot price or day-ahead price, system marginal price (SMP), locational marginal price (LMP) or zonal system marginal price (ZSMP) are implemented in different countries. Since, the main objective of market participants is to get maximum profit from the market, there is always scope of gaming, market power and market abuse. In the present competitive electricity market, various information are required in real sense for free and fair-trading of electricity along with reliable and secure operation of power system. The forecasting of various variables with high degree of accuracy is very essential in the emerging power system.

Load forecasting in a power system can normally be segregated into three categories: (i) short-term forecasting with a lead-time of up to a few days ahead, (ii) medium-term forecasting over a six month or one year period and (iii) long-term forecasting of the power system. A precise short-term load forecasting method plays a key role in the economic and secure operation of the power systems which allow electric utilities to optimize resources for better energy prices, reliable system operation, minimum operational costs and enhanced electricity market efficiency. The long- and medium- term load forecasting are used to determine the capacity of generation, expansion of transmission or distribution system and the type of facilities required in transmission expansion planning, annual hydro thermal maintenance scheduling, etc.

Many activities in the competitive electricity markets, such as trading and risk management, are directly dependent on the quality of the price forecasting in order to evaluate derivatives and devise hedging strategies. Accurate price forecasting helps utilities, independent power producers and customers to submit effective bids with low risks in order to maximize their benefits. Knowing their own costs, technical constraints and their anticipation of rival and market behaviour, the suppliers can forecast the best optimal bid of rivals and accordingly build their own bid, which will lead to more profit. Spinning reserve (SR) is one of the most important ancillary services (ASs) required for maintaining power system reliability following a major contingency [2]. An accurate short-term predication of day-ahead SR requirement helps the independent system operator (ISO) to make effective and timely decisions in managing the compliance and reliability of the power system.

In general, load, price, SR and bid in the competitive electricity markets are mutually intertwined activities. Load forecast, which is a major factor in the bid and price forecasting, is vital for the energy transactions in the competitive electricity markets. Errors in load forecasts and load scheduling are the major reasons of price swings. Forecasted load and price form the basis for bid forecasting. Accurately forecasted day-ahead SR requirements information help the market participants to develop an optimal bidding strategy that would maximize their profits in SR market. Thus, in present days, load, price, bid and SR forecasting have been central and integral process in the planning and operation of electric utilities.

3.2   Forecasting Methods

A range of methods for the load, price, bid and SR forecasting has been suggested in the literature. Some conventional methods of forecasting are time-of-day method, regression method, stochastic time series methods, state-space methods like auto-regressive moving average (ARMA), integrated auto-regressive moving average (ARIMA), box-Jenkins method, linear time series models, multivariate adaptive regression splines (MARS), generalized auto-regressive conditional hetero-scedasticity (GARCH) [3]. These forecasting approaches are having their own limitations in predicting the non-stationary, highly volatile signals. Artificial intelligence (AI) techniques such as fuzzy logic (FL), expert system, evolutionary computation (EC), genetic algorithm (GA), ant colony search (ACS), simulated annealing (SA), Tabu search (TS), particle swarm optimization (PSO) and artificial neural network (ANN) promise a good predictability or nearly so and have been emerged in recent years in power systems' variables forecasting. These tools provide better alternative compared to conventional methods for forecasting problems in power system area. However, these methods have main limitation of their sensitivity to the choice of parameters, such as the crossover and mutation probabilities in GA, temperature in SA, scaling factor in EP and inertia weight and learning factors in PSO [4]. However, AI relies heavily on good problem description and extensive domain knowledge.

Many researchers have applied wavelet transformation as a pre-processor to decompose the ill-behaved series into better behaved consecutive series and then forecasting models like ARIMA and ANN have been applied for forecasting [5]. To take care of the high-frequency changes, fuzzy model has been applied to forecast the possible ranges of variation in the electricity load and price. Dynamic fuzzy system (DFS), extended Kalman filters (EKF) and an input/output hidden Markov model (IOHMM) have been applied for the forecasting of variables in electricity markets [6]. A probabilistic methodology based on the integration of the loss of load cost (LOLC) concept in the capacity bidding process has been used to determine the operating reserve requirements and pricing.

ANN can acquire knowledge through adaptive training and generalization, hence, is the most useful AI technique for load, price, bid and SR forecasting. A feed forward multi-layer perceptron (MLP) model with back propagation (BP) training algorithms (gradient descent/ conjugate gradient method) has been applied for electricity load, price and bid forecasting. Cascaded architecture of multiple ANNs and committee machine replace the single neural network for complex nonlinear mapping functions. Radial basis function neural network (RBFNN) and recurrent neural network (RNN) have also been proposed for forecasting due to several advantages. A new fuzzy-neural network (FNN) technique with higher learning capability has been proposed to forecast market parameters [7].

3.2.1 *Load Forecasting Methods*

Many conventional methods such as time-series method, regression method, stochastic time series methods, state space methods [8-14] have been used for load forecasting. Expert system models are usually able to take both quantitative and qualitative factors into account. Many models of this type have been proposed since mid1980's. A typical approach is to imitate the reasoning of a human operator and to reduce the analogical thinking behind the intuitive forecasting to formal steps of logic [15]. A possible method for a human expert to forecast is to search in historical database for a day that corresponds to the target day with factors like day type, social factors and weather factors. Then the load values of this similar day are taken as the basis for the forecast. An expert system can thereby be an automated version of this kind of a search process [16]. On the other hand, the expert system can consist of a rule base defining relationships between external factors and daily load shapes. A popular approach to develop rules on the basis of fuzzy logic has been reported in reference [17]. The heuristic approach in arriving at solutions makes the expert system methods attractive for the system operators [18].

In recent years, artificial neural networks (ANNs) have been applied to many areas of power system analysis and control such as load forecasting [19], static and dynamic security assessment, dynamic load modelling, alarm processing and fault diagnosis [20].

These applications take advantage of the powerful mapping ability of ANNs and their inherently parallel and distributed processing characteristics for performing ultra-high-speed computation. Artificial neural network (ANN), whose operation is based on certain known properties of biological neurons, comprises various architectures of highly interconnected processing elements that offer an alternative to conventional computing approaches. They can achieve complicated input-output mappings without explicit programming and extract relationships (both linear and nonlinear) among data sets presented during a learning process. Furthermore, the redundancy of their inter connections ensures robustness and fault tolerance and they can be designed to self-adapt and learn [21], [22].

The application of ANNs to the short-term load forecasting has gained a lot of attention. Dillon et al. [23] used adaptive pattern recognition and self-organizing techniques for short term load forecasting. Later, they used an adaptive neural network for short term load forecasting [24]. The availability of historical load data in the utility databases makes this area highly suitable for ANN implementation. ANNs are able to learn the relationship among weather variables and loads. As in the time series approach, the ANN traces previous load patterns and predicts (i.e., extrapolates) a load pattern using recent load data [25]. The ANN is able to perform non-linear modelling and adaptation. Their ability to perform better than traditional methods especially during rapidly changing weather conditions and the less time, have made ANN based load forecasting models very attractive for on-line implementation in energy control centres.

3.2.2 *Price Forecasting Methods*

Market simulation methods, which, generally, utilize several hypotheses, consider the generators' operation cost with the generator and transmission constraints. This approach is suitable for long-term price forecasting. Statistical methodology, which is based on the assumption of historical price characteristics, can be categorized in to the time series models, intelligent system methods and volatility analysis. Many of these methods are used for the short term load forecasting (STLF) [26].

Time-series models provide a trade-off between underlying price behaviour and accurate forecasting. Contreras et al. [27] have developed auto regressive integrated moving average (ARIMA) model to forecast electricity market prices of the Spanish and Californian markets. Multivariate dynamic regression (DR) and transfer function (TF) models have been applied to forecast the Spanish and California market prices [28] and the PJM market prices [29], [30]. DR relates current price to the past prices and past demands whereas TF relates current price to past prices, past demands and past errors. Time series models have been also applied to forecast commodity prices [31] such as oil [32] and natural gas [33]. Early applications of the time series models in the power system were for STLF [14]. Simple auto regressive (AR) models are also being used to predict weekly prices in the Norwegian system [34]. A Bayesian-based classification method combined with an AR model is presented in reference [35] to predict the discrete probability density functions (PDF) of the market clearing prices (MCP). In reference [36], the performance of ARIMA model was improved by incorporating forecasted error.

The non-linear multivariate adaptive regression splines (MARS) technique has been applied to forecast hourly energy price [37]. Generalized auto-regressive conditional hetero-scedasticity (GARCH) uses past variances and past variance forecasts to forecast future variance and has been applied to forecast day-ahead electricity price for Spain and California market [38], [39].

A typical artificial neural network (ANN) for electricity price forecasting is feed forward multi-layer perceptron (MLP) model with back propagation (BP) training algorithm (gradient descent) [40], [41] or with conjugate gradient algorithm [42]. A single neural network with traditional learning algorithms may not be suitable for complex nonlinear mapping function of the price signal. Cascaded architecture of multiple ANNs [43] and committee machine [44] replace the single neural network. Radial basis function neural network (RBFNN) [45] and recurrent neural network (RNN) [46] have also been proposed for forecasting due to several advantages.

To take care of the high-frequency changes of the MCP, fuzzy model [47] has been applied to forecast the possible ranges of variation in the electricity price. L. Hongjie et al. [48] have proposed dynamic fuzzy system (DFS) to forecast MCP of California power market. A kind of extended Kalman filter (EKF) [49] and an input/output hidden markov model (IOHMM) [50] have been applied to New England and Spain electricity markets, respectively. A new fuzzy NN (FNN) with higher learning capability than ANNs has been proposed to forecast electricity prices [51]. Many researchers have applied wavelet transformation as a preprocessor to decompose the ill-behaved price series into better behave consecutive series and then forecasting model like ARIMA [52] and ANN [53] have been applied for price forecasting.

Simulation methods can provide detail insights into the price curve. Market assessment and portfolio strategies (MAPS) model, which is based on locational marginal price (LMP), is a transmission-constrained chronological production simulation model [51]. A structural multi-commodity multi-area optimal power flow (MMOPF) is also a product simulation model that performs Monte Carlo simulation to take into account all the major drivers, including participant's bidding behaviour, can provide reliable prices and realistic option values.

Different approaches for measuring wholesale electricity market price volatility have been proposed. Alvarado et al. [54] have modelled electricity prices using a frequency domain analysis to separate the cyclic component of price. A study by R. Deb et al. [55] shows that in Spain, California, and PJM electricity markets, the price volatility is strongly connected to the installed generation capacity. T. Mount [56] argued that the use of a uniform price auction for electricity markets worsens price volatility and a pay-as-bid price auction is a better alternative.

In the electricity market, the price of the electricity is no more defined by the monopolist, who operated the power industry before the deregulation. Restructuring of electricity supply industry, worldwide, has brought a variety of new issues such as oligopolistic nature of the market, supplier's strategic bidding, market power abuses, price-demand elasticity and so on. The suppliers bid in the market with various strategies to maximize their profit and therefore, determine the market behaviours. Many researchers have studied the bidding strategies and their impacts on the market behaviour. Game theory, which has been widely applied for analysing the behaviour of the market players [57], [58] is a useful tool to study the interactions among the players. But game theory application needs a lot of detailed information about the market. It is difficult to capture the exact situation of the market and other participants. A small change even in the mind of one participant may possibly results in huge difference in market result, hence game theory is good for theoretically explaining an electricity market and providing a reasonable market behaviour, but not for predicting the exact market outcome.

Theoretically, in a perfectly competitive market, suppliers should bid at, or very close to, their marginal production cost to maximise the payoff [59]. Also producers, who are small enough to affect market prices with their bids, are price takers, and their optimal strategy is to bid at the marginal cost of production [60]. However, the electricity markets are oligopolistic in practice, and power suppliers may seek to increase their profit by bidding a price higher than marginal production cost. Knowing their own costs, technical constraints and their anticipation of rival and market behaviour, suppliers construct the best optimal bid. There are three main approaches to model the strategic bidding problems based on the estimation of market clearing price, estimations of rival's bidding behaviour and game theory. In the recent years, a considerable amount of work has been published on strategic bidding [61] in electricity markets. David [62] proposed a conceptual optimal bidding model for the first time in which a dynamic programming based (DP) approach has been used to solve strategic bidding problem. A lagrangian relaxation-based approach for strategic bidding in England-Wales pool type electricity market has been adopted [63]. The same approach for daily bidding and self-scheduling decision in New England market has been suggested by Zhang et al. [64].

Considerable work has also been reported on the game theory applications in the competitive electricity markets. In non-cooperative game theory approach [65], [66] strategic bidding problem is solved using Nash equilibrium. Genetic algorithm (GA) has been proposed by David and Wen [67] to develop an overall bidding strategy using two different bidding schemes for a day-ahead market. The same methodology has been extended for spinning reserve market coordinated with energy market [68]. Ugedo et al. [69] have proposed a stochastic-optimization approach for submitting the block bids in sequential energy and ancillary services markets, and uncertainty in demand and rival's bidding behaviour is estimated by stochastic residual demand curves based on the decision trees. A stochastic programming model has been used to construct linear bid curves in the Nord-pool market for price-taking retailer whose customers' load is price flexible [70]. Profit and volume deviation risks have been incorporated in the model for single hourly trading in the day-ahead and short-term balancing markets. Opponents' bidding behaviours are represented as a discrete probability distribution function [71] and as a continuous probability distribution function [72] for a supplier's bid decision-making problem.

Strategic bidding problem has been formulated as a two-level optimization problem in reference [73]–[77]. The producers try to maximise their profits by optimally bidding in the market. Using deterministic approach, it is difficult to obtain the global solution of such bi-level optimization problem because of non-convex objective function and non-linear functions [73], [74] to represent the market clearing. These difficulties are avoided by representing the residual demand function by mixed integer linear programming (MILP) model [75], [76] in which unit commitment and uncertainties are also taken in to account. The generators associated to the competitors' firms explicitly models an alternative MILP formulation based on a binary expansion of the decision variables (price and quantity bids) [77].

The strategic bidding problem can be solved by various conventional and non-conventional (heuristic) methods such as GA, simulated annealing (SA) and evolutionary programming (EP), particle swarm optimization (PSO), etc. However these methods have main limitation of their sensitivity to the choice of parameters. PSO is a modern stochastic search algorithm and a kind of evolutionary computation technique. The PSO technique can generate better quality solution within shorter calculation time and stable convergence characteristic than other stochastic methods [78]-[80]. Two approaches for forming the market bidding strategies based on SVM are proposed in reference [81]. One is based on the price forecast accuracy and other takes into account the impact of the producer's own bid.

3.2.3 *Spinning Reserve Forecasting Methods*

The spinning reserve (SR), which is the fastest-responding contingency reserve, is the most critical for maintaining power system reliability following a major contingency, such as the unplanned loss of a large generation facility or outage of critical transmission line. Several system operators (SOs) have adopted deterministic criteria to access the spinning reserve requirements. According to their operating rules, the spinning reserve should be greater than the capacity of the largest online unit or a fraction of the load, or equal to some combination of both [82]. The California ISO (CAISO) ensures that at least half of operating reserves must be SR for each settlement period of the market. As per the deterministic criteria adopted by CAISO, the operating reserve requirement is equal to 5% of the load to be supplied by hydroelectric resources, plus 7% of the load to be supplied by generation from other resources, plus 100% of any interruptible imports, or a single largest contingency [83]. The deterministic criteria, however, do not reflect the uncertainty in the forecast load and unpredicted outage of generating units. The CAISO procures required operating reserves (which are decided based on deterministic criteria) through competitive bidding process.

The SO's objective is to minimize AS payments while encouraging participants to provide sufficient AS. On the other side, energy producers would anticipate submitting a bid that would maximize their profits in the market [84]. Probabilistic techniques [85] have been proposed for computing SR requirements in a more consistent way. A probabilistic methodology based on the integration of the loss of load cost (LOLC) concept in the capacity bidding process has been presented in reference [86], to determine the operating reserve requirements and pricing. Artificial neural network (ANN) has been applied for SR probability forecasting in reference [87].

3.3   Summary

In the competitive electricity markets, there are several key variables which extremely influence the market operation. These variables such as load, electric price, bids of rival suppliers, spinning reserves, etc. are dependent on the several factors and very difficult to predict accurately. Moreover, these variables are correlated and suffer from the gaming of the market participants. Several approaches using the conventional methods are suggested to predict these variables but these conventional methods suffer from the accuracy and capturing the volatility of the variables. The need for fast answer to complicated problems having uncertain and incomplete information/data undoubtedly grows as deregulation progresses [22]. This situation presents an enormous opportunity for AI applications to solve the challenging problems looming in the future of power systems. Artificial intelligence (AI) tools have been useful for solving power system problems when there is a good match between the problem characteristics and those of the artificial intelligence (AI) tool.

References

[1] K. G. Upadhyay, "Design of lectric utility restructuring model for   Indian Power System", Ph.D. Thesis, U. P. Technical University, Lucknow, India, October 2001.

[2] N. M. Pindoriya, S. N. Singh, S. K. Singh, "Forecasting the Day-Ahead Spinning Reserve Requirement in Competitive Electricity Market", IEEE Power and Energy Society General

Meeting - Conversion and Delivery of Electrical Energy in the 21st Century, July 2008.

[3] http://www.wacong.org/autosoft/auto/documents/mmt_pjm.pdf

[4] Kumar, J. Vijaya and Kumar, D. M. Vinod, "Congestion Influence on Optimal Bidding in a Competitive Electricity Market Using Particle Swarm Optimization", Majlesi Journal of Electrical Engineering, 2011.

[5] J. P. S. Catalão, H. M. I. Pousinho, V. M. F. Mendes, "Hybrid Wavelet-PSO-ANFIS Approach for Short-Term Electricity Prices Forecasting" IEEE TRANSACTIONS ON POWER SYSTEMS, http://webx.ubi.pt/~catalao/05471112.pdf

[6] Filipe Azevedo, Zita A. Vale, "Forecasting Electricity Prices with Historical Statistical Information using Neural Networks and Clustering Techniques", 142440178-X/06/$20.00 ©2006 IEEE, PSCE 2006, pp 44-50

[7] Sanjeev Kumar Aggarwal, Lalit Mohan Saini, Ashwani Kumar, "Electricity price forecasting in deregulated markets: A review and evaluation", Electrical Power and Energy Systems 31 (2009), pp 13–22

[8] Pauli Murto, "Neural network models for short-term load forecasting", Masters thesis, Department of Engineering Physics and Mathematics, Helisinki University of technology, Helsinki, January 5, 1998.

[9] M. Räsänen and J. Ruusunen, "Verkoston tilan seurantamittauksilla jakuormitusmalleilla", Research Report B17, Systems Analysis Laboratory, Helsinki University of Technology.

[10] A. D. Papalexopoulos and T. C. Hesterberg, "A regression-based approach to short-term system load forecasting", IEEE Trans. on Power Systems, Vol. 5, No. 4, November 1990, pp. 1535-1547.

[11] M. T. Hagan and S. M. Behr, "The time series approach to short term load forecasting", IEEE Trans. on Power Systems, Vol. PWRS-2, No. 3, August 1987, pp. 785-791.

[12] R. S. Pindyck and D. L. Rubinfeld, "Econometric models and economic forecasts", McGraw-Hill, Singapore.

[13] R. Campo and P. Ruiz, "Adaptive weather-sensitive short-term load forecast", IEEE Trans. on Power Systems, Vol. PWRS-2, No. 3, August 1987, pp. 592-600.

[14] G. Gross and F. D. Galiana, "Short-term forecasting", Proceedings of the IEEE, Vol. 75, No. 12, 1987, pp. 1558-73.

[15] S Rahman and R Bhatnagar, "An expert system based algorithm for short term load forecast", IEEE Trans. on Power Systems, Vol. 3, No. 2, May 1988, pp. 392-399.

[16] K. Jabbour, J. F. V. Riveros, D. Landsbergen and W. Meyer, ALFA, "Automated load

forecasting Assistant", IEEE Trans. on Power   Systems, Vol. 2, No. 3, August 1988, pp. 908-914.

[17]   Y. Y. Hsu and K. L. Ho, "Fuzzy expert systems: an application to short-term load forecasting", IEE Proceedings-C, Vol. 139, No. 6, November 1992, pp. 471-477.

[18] A. Asar and J. R. McDonald, "A specification of neural network   applications   in   the   load forecasting problem", IEEE Trans. on   Control   Systems   Technology, Vol.   2,   No.   2, November 1992, pp. 135-141.

[19]   D. C. Park, M. A. El-Sharakawi and R.I. Marks, "Electric load   forecasting using artificial neural networks", IEEE Trans. on Power Systems, Vol. 6,1991, pp. 442–449.

[20]   N. Kandil, V. K. Sood, K. Khorasani and R. V. Patel, "Fault identification in an AC–DC transmission system using neural networks", IEEE Trans. on Power Systems, Vol. 7, No. 2, 1992, pp. 812–819.

[21]   B. Soucek and M. Soucek, "Neural and massively parallel computers, the sixth generation", John Wiley, New York, USA, 1988.

[22]   B. Widrow and D. Stems, "Adaptive signal processing", Prentice- Hall,   New   York,   USA, 1985.

[23]   T. S. Dillon, K. Morsztyn and K. Phua, "Short term load forecasting using adaptive pattern recognition and self-organizing techniques", Proceedings of the fifth power system computation conference (PSCC05), Cambridge, paper 2.4/3, September 1975, pp. 1–15.

[24]   T. S. Dillon, S. Sestito and S. Leung, "Short term load forecasting using an adaptive neural network", International Journal of Electrical Power and Energy Systems, Vol. 13, 1991, pp. 186–92.

[25]   N. Kandil, R. Wamkeue, M. Saad and S. Georges, "An efficient   approach   for   short   term load forecasting using artificial neural   networks", International Journal of Electrical Power and Energy Systems, Vol. 28, 2006, pp. 525-530.

[26]   N.Venkata Rao, K.Sarada, "PRICE EASTIMATION FOR DAY-AHEAD ELECTRICITY MARKET USING FUZZY LOGIC", International Journal of Advanced Research in Electrical, Electronics and Instrumentation Engineering, Vol. 2, Issue 5, May 2013, pp 1940-1946

[27]   J. Contreras, R. Espínola, F. J. Nogales and A. J. Conejo, "ARIMA Models to Predict Next-Day Electricity Prices," IEEE Trans. on Power Systems, Vol. 18, No. 3, August 2003, pp. 1014–1020.

[28]   F. J. Nogales, J. Contreras, A. J. Conejo and R. Espínola, "Forecasting Next-Day Electricity Prices by Time Series Models," IEEE Trans. on Power Systems, Vol. 17, No. 2, May 2002, pp. 342-348.

[29] A. J. Conejo, J. Contrearas, R. Espinola and M. A. Plazas, "Forecasting Electricity Prices for a day ahead pool based electric energy market," International Journal of Forecasting, Vol. 21, 2005, pp. 435–462.

[30] H. Zareipour, K. Bhattacharya, C.A. Canizares, "Forecasting the Hourly Ontario Energy Price by Multivariate Adaptive Regression Splines", http://thunderbox.uwaterloo.ca/ ~ccanizar /papers/ HamidMontreal.pdf

[31] E. Weiss, "Forecasting commodity prices using ARIMA," Technical Analysis of Stocks & Commodities, Vol. 18, No. 1, 2000, pp. 18–19.

[32] C. Morana, "A semiparametric approach to short-term oil price forecasting," Energy Economics, Vol. 23, No. 3, May 2001, pp. 325– 338.

[33] W. K. Buchananan, P. Hodges and J. Theis, "Which way the natural gas price: An attempt to predict the direction of natural gas spot price movements using trader positions," Energy Economics, Vol. 23, No. 3, May 2001, pp. 279–293.

[34] O. B. Fosso, A. Gjelsvik, A. Haugstad, M. Birger and I. Wangensteen, "Generation scheduling in a deregulated system. The norwegian case," IEEE Trans. Power Systems, Vol. 14, No. 1, Feb.   1999, pp. 75–81.

[35] E. Ni and P. B. Luh, "Forecasting Power Market Clearing Price and its Discrete PDF using a Bayesian-based Classification Method," IEEE Power Engineering Society Winter Meeting, Vol. 3, Columbus, OH, 2001, pp. 1518–1523.

[36] M. Zhou, Z. Yan, Y. Ni and G. Li, "An ARIMA approach to forecasting electricity price with accuracy improvement by predicted errors," Proc. IEEE Power Engineering Society General Meeting, June 2004, pp. 233–238.

[37] H. Zareipour, K. Bhattacharya and C. A. Ca~nizares, " Forecasting the Hourly Ontario Energy Price by Multivariate Adaptive Regression Splines", Proc. IEEE Power Engineering Society General Meeting, Montreal, Quebec, Canada, 18–22 June, 2006.

[38] R. C. Garcia, J. Contreras, M. van Akkeren and J. B. C. Garcia, "A GARCH Forecasting Model to Predict Day-Ahead Electricity Prices," IEEE Trans. on Power Systems, Vol. 20, No. 2, May 2005, pp. 867 – 874.

[39] Hu Linlin, Taylor GA, "A novel hybrid technique for short-term electricity price forecasting in deregulated electricity markets", Brunel University School of Engineering and Design PhD Theses, 2010

[40] P. Doulai and W. Cahill, "Short Term Price Forecasting in Electric Energy market", International Power Engineering Conference (IPEC 2001), May 2001, pp. 749–754.

[41]   P. Mandal, T. Senjyu, K. Uezato and T. Funabashi, "Several Hours Ahead Electricity Price and Load Forecasting Using Neural Networks ", IEEE Power Engineering Society General Meeting, Vol. 3, 12-16 June 2005, pp.   2146–2153.

[42]   A. Sedaghati, "Using Neural Network to Forecast Price in Competitive Power Market", International Conference on Control,   Automation & System (ICCAS), Korea, June 2–5, 2005.

[43]   L. Zhang, P. B. Luh and K. Kasiviswanathan, "Energy Clearing Price Prediction and Confidence Interval Estimation With Cascaded Neural Networks," IEEE Trans. on Power Systems, Vol. 18, No. 1, Feb. 2003, pp. 99–105.

[44]   J. J. Guo and P. B. Luh, "Improving Market Clearing Price Prediction by using a Committee Machine of Neural Networks," IEEE Trans. on Power Systems, Vol. 19, No. 4, Nov. 2004, pp. 1867–1876.

[45]   B. R. Szkuta, L. A. Sanabria and T. S. Dillon, "Forecasting Power Market   Clearing   Prices and Quantity Using Neural Network Method", IEEE Power Engineering Society Summer Meeting, 2000, pp. 2183–2188.

[46]   F. Gao, X. Guan, X. Cao and A. Papalexopoulos, "Forecasting Power Market Clearing Price and Quantity Using a Neural Network Method," Proc. of Power Engineering Summer Meeting, Seattle, WA, July 2000, Vol. 4, pp. 2183–2188.

[47]   T. Niimura and T. Nakashima, "Deregulated electricity market data representation by fuzzy regression models," IEEE Trans. on Systems, Man and Sybernetics–Part c: Applications and Reviews, Vol. 31, No. 3, Aug. 2001, pp. 320–326.

[48]   L. Hongjie, W. Xiugeng, Z. Weicun and X. Guohua, "Market   clearing price forecasting based on dynamic fuzzy system," Proc. PowerCon 2002, International Conference on Power System Technology, Vol. 2, 13–17 Oct. 2002,   pp. 890–896.

[49]   L. Zhang and P. B. Luh, "Neural network-based market clearing   price   prediction   and confidence interval estimation with an   improved extended Kalman filter method," IEEE Trans. on Power Systems, Vol. 20, No. 1, Feb. 2005, pp. 59–66.

[50]   A. M. Gonzalez, A. M. S. Roque, and J. Garcia-Gonzalez, "Modelling and Forecasting Electricity Prices with Input/Output Hidden Markov Models," IEEE Trans. on Power Systems, Vol. 20, No. 1, Feb. 2005, pp. 13–24.

[51]   N. Amjady, "Day-Ahead Price Forecasting of Electricity Markets by a New Fuzzy Neural Network", IEEE Trans. on Power System, Vol. 21, No.2, May 2006, pp.887–896.

[52]   H. Xu and T. Niimura, "Short-Term Electricity Price Modeling and Forecasting   Using Wavelets and Multivariate Time Series," IEEE Power Engineering Society General Meeting, Vol.

1, 10–13 Oct. 2004, pp. 208 – 212.

[53]   V. Iyer, C. C. Fung and Tamas, "A Fuzzy-Neural Approach to   Electricity Load and Spot price Forecasting in a Deregulated Electricity Market", IEEE Conference on Convergence Technology for Asia–Pacific Region, Vol. 4, 15–17 Oct. 2003, pp. 1479–1482.

[54]   F. L. Alvarado and R. Rajaraman, "Understanding Price Volatility in Electricity   Markets," Proc. of the 33rd Hawaii International   Conference on System Sciences, 4-7 Jan. 2000.

[55]   R. Deb, R. Albert, L.-L. Hsue and N. Brown, "How to Incorporate Volatility and Risk in Electricity Price Forecasting," The Electricity   Journal, May 2000.

[56]   T. Mount, "Market power and price volatility in restructured markets for electricity", Decision Support Systems, Vol. 30, 2001, pp. 311–325.

[57]   D. Chattopadhyay, "Multicommodity spatial Cournot model for   generatorbidding analysis", IEEE Trans. on Power Systems, Vol. 19, No. 1, 2004, pp. 267 – 275.

[58]   S. De la Torre, J. Contreras and A. J. Conejo, "Finding multiperiod Nash   equilibria in pool-based electricity markets", IEEE Trans. on Power Systems, Vol. 19, No. 1, 2004, pp. 643 – 651.

[59]   Kumar, J. Vijaya and Kumar, D. M. Vinod, "A Literature Review on Strategic Bidding in an Open Access Electricity Market", International Review on Modelling & Simulations, 2012.

[60]   G. Gross and D. Finlay, "Generation supply bidding in perfectly   competitive electricity markets", Computational & Mathematical Organization Theory, Vol. 6, No. 1, May 2000, pp. 83–98.

[61]   A. K. David and F. Wen, "Strategic bidding in competitive electricity markets: a literature survey", IEEE PES Summer Meeting, 2000, Vol. 4, pp. 2168–   2173.

[62]   A. K. David, "Competitive bidding in electricity supply", IEE Proc. Generation, Transmission and Distribution, 1993, Vol. 140, No. 5, pp. 421–426.

[63]   G. Gross and D. J. Finlay, "Optimal bidding strategies in competitive electricity markets", Proc. 12th Power System Computation Conference, August 1996, pp. 815–823.

[64]   D. Zhang, Y. Wang and P. B. Luh, "Optimization based bidding   strategies   in   deregulated market", Proc. IEEE PES Power Industry Computer Applications Conference, 1999, pp. 63–68.

[65]   R. W. Ferrero, V. C. Ramesh and S. M. Shahidehpour, "Transaction   analysis   in deregulated power system using game theory", IEEE   Trans. Power Systems, Vol. 12, No. 3, 1997, pp. 1340–1347.

[66]   S. D. Torre, A. J. Conejo and J. Contreras, "Finding multi-period Nash   equilibrium   in pool-based electricity markets", IEEE Trans. Power Systems, Vol. 19, No. 1, 2004, pp. 643–651.

[67]   A. K. David and F. Wen, "Strategic bidding for electricity supply in a   day-ahead   energy

market", Electric Power System Research, Vol. 59, 2001, pp. 197–206.

[68]   A. K. David and F. Wen, "Optimally co-ordinated bidding strategies   in   energy   and ancillary service markets", IEE Proc. Generation, Transmission and Distribution, Vol. 149, No. 3, 2002, pp. 331–338.

[69]   A. Ugedo, E. Lobato, A. Franco, L. Rouco, J. Fernaˊndez-Caro and J. Chofr, "Strategic bidding in sequential electricity markets", IEE   Proc. Generation, Transmission and Distribution, Vol. 153, No. 4, 2006, pp. 431–442.

[70]   S. E. Fleten and E. Pettersen, "Constructing bidding curves for a   price-taking retailer in the Norwegian electricity market", IEEE Trans. Power Systems, Vol. 20, No. 2, 2005, pp. 701–708.

[71]   H. L. Song, C. C. Song and J. Lawree, "Decision making of an   electricity suppliers bid in a spot market", Proc. IEEE Power Engineering Society Summer Meeting, Vol. 1, 1999, pp. 692–696.

[72]   A. K. David and F. Wen, "Optimal bidding strategies and modeling of imperfect information among competitive generators", IEEE   Trans. Power Systems, Vol. 16, No. 1, 2001, pp. 15–21.

[73]   B. F. Hobbs, C. B. Metzler and J. S. Pang, "Strategy gaming analysis for electric power systems: an MPEC approach", IEEE Trans. Power Systems, Vol. 15, No. 2, 2000, pp. 638–645.

[74]   B. F. Hobbs and U. Helman, "Complementarity-based equilibrium modeling for electric power markets", in Bunn, D. (Ed.): 'Modeling prices in competitive electricity markets' Wiley, New York, 2004.

[75]   A. Baillo, M. Ventosa, M. Ventosa and A. Ramos, "Optimal offering strategies for generation companies operating in electricity spot markets", IEEE Trans. Power Systems, Vol. 19, No. 2, 2004, pp. 745–753.

[76]   S. Torre, J. M. Arroyo, A. J. Conejo and J. Contreras, "Price maker self-scheduling in a pool-based electricity market: a mixed-integer LP approach", IEEE Trans. Power Systems, Vol. 17, No. 4, 2002, pp. 1037–1042.

[77]   M. V. Pereira, S. Granville, M. Fampa, R. Dix and L. A. Barroso, "Strategic bidding under uncertainty: a binary expansion approach", IEEE Trans. Power Systems, Vol. 20, No. 1, 2005, pp. 180–188.

[78]   J. Kennedy and R. Eberhart, "Particle swarm optimisation'. Proc. IEEE   International Conference on Neural Networks" Vol. 4, 1995, pp. 1942–1948.

[79]   P. Bajpai, S.K. Punna and S.N. Singh, "Swarm intelligence-based strategic bidding in competitive electricity markets", IET Generation, Transmission and   Distribution, Vol. 2, No. 2, 2008, pp. 175–184.

[80]   H. Shayeghi, H. A. Shayanfar, G. Azimi, "STLF Based on Optimized Neural Network Using PSO", International Scholarly and Scientific Research & Innovation, World Academy of Science, Engineering and Technology Vol. 3, 2009, pp 874-884

[81]   C. Gao1, E. Bompard, R. Napoli and Q. Wan, "Bidding strategy with   forecast technology based on support vector machine in electricity   market", Applications of Physics in Financial Analysis 6th International Conference (APFA-6), Lisbon, Portugal, 4-7 July 2007.

[82] H. B. Gooi, D. P. Mendes, K. R. W. Bell and D. S. Kirschen, "Optimal scheduling of spinning reserve", IEEE Trans. Power   Systems, Vol. 14, No. 4, Nov. 1999, pp. 1485-1492.

[83]   A report on "Ancillary services capacity settlement in CAISO controlled grid," Available at http://www.caiso.com.

[84] M. Shahidehpour, H. Yamin and Z. Li, "Market operations in electric power systems: forecasting, scheduling, and risk management", New   York: Wiley, 2002.

[85] L. T. Anstine, R. E. Burke, J. E. Casey, R. Holgate, R. S. John and H. G. Stewart, "Application of probability methods to the determination of spinning reserve requirements for the Pennsylvania - New Jersey - Maryland interconnection", IEEE Trans. PAS, PAS-82, Oct. 1963, pp. 726–735.

[86]   A. M. Leite da Silva and G. P. Alvarez, "Operating reserve capacity requirements and pricing in deregulated markets using probabilistic techniques", IET Generation, Transmission and Distribution, Vol. 1, No. 3, 2007, pp. 439–446.

[87] H. Y. Yamin, "Spinning reserve uncertainty in day-ahead competitive electricity markets for GENCOs", IEEE Trans. Power Systems, Vol. 20, No. 1, Feb. 2005, pp. 521-523.

[88]   Tahir Nadeem Malik, Aftab Ahmad, "Potential of Artificial Intelligence (AI) Techniques for WAPDA Grid System in Deregulated Environment", National Conference on Emerging Technologies 2004, pp 58-62

CHAPTER 4

Load and Price Forecasting Methods

4.1   Introduction

Present day power systems are changing from old monopolistic structure to vertically unbundled structure to introduce competition in the electricity supply industry. In the new environment, power industry's main challenges are to enhance the economic and operational efficiencies, reliability and stability. Although the fundamental technologies of power generation, transmission and distribution change quite slowly, the power industry is changing in terms of operation, ownership, structure and management. Thus, power system becomes highly man made, large and complex, and has several potential issues which are to be solved effectively and timely. Time scales involved for various activities in power system planning, operation and control vary from seconds to minutes, hours, months and even years.

Load forecasting, price and bid forecasting have been central and integral part in the planning, operation and control of power systems. Spinning reserve (SR) forecasting has also become important in competitive power market for stable and secure operation of the electric power system. Many operating decisions such as dispatch scheduling of generating capacity, reliability analysis and maintenance scheduling of the generating units are based on load forecasts. Load forecasts are also vital for the energy transactions in competitive electricity markets. In addition, the accurate estimated load is key data for the electricity price and bid forecasting. Electricity price is useful to the market participants for building their optimal bidding strategies in a deregulated power market and suitable price forecasting will help utilities and independent power producers to submit effective bids with low risks. In the competitive electricity markets, formation of supply bid is one of the main concerns for generators where suppliers have to maximize their profit under incomplete information of other competing generators.

Mathematical optimization (algorithmic) methods have been used over the years for many power systems planning, operation and control problems [1]. Mathematical formulations of real-world problems are derived under certain assumptions and even with these assumptions, to get a feasible optimal solution of large-scale power systems is difficult. On the other hand, there are many uncertainties in power system problems because power systems are large, complex and geographically widely distributed. More recently, deregulation of power utilities has introduced some new issues into the existing problems. It is desirable that solution of power system problems should be optimum globally, but solution searched by mathematical optimization is normally optimum local.

The researchers in the power engineering community have been studying the feasibility and application of new information processing techniques for efficient solution of complex power system problems. With the development of artificial intelligence (AI) tools, alternative methods have been identified, proposed and developed for the solution of ill-structured and complex problems in power systems [2]. This chapter presents an overview of conventional methods and important artificial intelligence (AI) methods used in load, price, bid and SR forecasting.

4.2   Conventional Methods

Many conventional methods have been used to forecast the electricity market variables [2], which have neither any mathematical model nor explicit function of power system parameters. Some of these methods are discussed below.

4.2.1   *Time-of-day Method*

A time-of-day method takes the previous week's actual load pattern to predict the load of the present week. Alternatively, a set of load patterns is stored for typical weeks with different weather conditions. These are, then, heuristically combined to create the forecast. Commonly, a time-of-day method is of the form (4.1).

$$Z(t) = \sum_{i=1}^{N} \alpha_i f_i(t) + v(t) \tag{4.1}$$

where the load at time t is expressed as a weighted sum of explicit time functions, usually sinusoids with a period of 24 or 168. The coefficients α_i are slowly varying constants, being usually estimated through a linear regression or exponential smoothing. The modeling error $v(t)$ is assumed to be white noise. $Z(t)$ is output function and $f_i(t)$ is input function.

4.2.2 Regression Method

Regression method, normally, assumes that the load can be divided into a standard load component and a component linearly dependent on some explanatory variables. This can be written as:

$$Z(t) = b(t) + \sum_{i=1}^{N} a_i \times y_i(t) + v \times \varepsilon(t) \tag{4.2}$$

where $b(t)$ is the standard load, $\varepsilon(t)$ is a white noise component and $y_i(t)$ are the independent explanatory variables at time t. The most typical explanatory variables are weather factors. A typical regression method has been used by Räsänen and Ruusunen [4]. In this method, the load is divided into a rhythm component that corresponds to the load of a certain hour in the average temperature of the modeling period and a temperature dependent component. Some models use previous load values as explanatory variables in addition to the external variables [5]. Regression methods are quite insensitive to occasional disturbances in the measurements. The easy implementation is the strength of this method. The serial correlation, which is typical when regression models are used in time series, can cause problems.

4.2.3 Stochastic Time Series Methods

This is a very popular dynamic forecasting method [6]. There are many types of stochastic time series methods such as auto-regressive moving average (ARMA), integrated auto-regressive moving average (ARIMA), box-Jenkins method, linear time series methods, etc. A general treatment of these methods can be found in reference [7]. The basic principle is that the load time series can first be transformed into a stationary time series by a suitable differencing approach. Then, the remaining stationary series can be filtered into white noise. The method assumes that the properties of the time series remain unchanged for the period used in model estimation and all disturbances are due to the white noise component. The basic ARIMA model can be written as:

$$\phi(B)\nabla^d Z(t) = \theta(B)a(t) \tag{4.3}$$

where $Z(t)$ and $a(t)$ are the modeled time series and a white noise sequence for $t = 1, \ldots, N$, respectively. $\phi(B)$ is the auto-regressive (AR) parameter polynomial and $\theta(B)$ is the moving average (MA) parameter polynomial defined as:

$$\phi(B) = 1 - \phi_1 B - \ldots\ldots\ldots - \phi_p B^p \tag{4.4}$$

$$\theta(B) = 1 - \theta B - \ldots\ldots\ldots - \theta_q B^q \tag{4.5}$$

where B is the backward shift operator $(B^n(z(t)) = z(t-n))$, ϕ_i, θ_i are constant parameters and $\nabla = 1 - B$ is a backward difference operator.

This basic ARIMA method is not by itself suitable for describing the load time series, since the load series incorporates seasonal variation. Therefore, the differencing with the period of seasonal variation is required. The method obtained is called a seasonal ARIMA (SARIMA) method and can be written as:

$$\phi(B)\phi_s(B^s)\nabla^d\nabla_s^D Z(t) = \theta(B)\theta_s(B^s)a(t) \tag{4.6}$$

where $\nabla_s^D = (1 - B^s)^D$ and s is the period of the seasonal variation.

An external input variable, such as temperature in the case of load time series, can also be included in the model. Such a variant of the ARIMA method is called an ARIMAX method and can, in general, be written as:

$$\phi(B)\nabla^d Z(t) = w(B)x(t-b) + \theta(B)a(t) \tag{4.7}$$

where, $x(t)$ is the external variable at time t and

$$w(B) = w_0 + w_1 B + w_2 B^2 + w_3 B^3 + \ldots\ldots \tag{4.8}$$

The ARIMA method including both the seasonal variation and external variable is, sometimes, called a SARIMAX method.

The stochastic time series methods have many attractive features. First, the theory of the methods is well known and therefore, it is easy to understand how the forecast is composed. The properties of the model are easy to calculate and the estimate for the variance of the white noise component allows the confidence intervals for the forecasts to be created. Established methods for diagnostic checks are also available. The weakness in the stochastic methods is in its adaptability. A forgetting factor can be used to give more weight to the most recent behaviour and thereby improves the adaptability. Another problem is the handling of the anomalous load conditions. If the load behaviour is abnormal on a certain day, this deviation from the normal conditions will be reflected in the future forecasts.

4.2.4 *State Space Method*

The linear state-space method can be written as:

$$Z(t) = c^T x(t) \tag{4.9}$$

where $x(t+1) = Ax(t) + Bu(t) + w(t)$

The state vector at time t is $x(t)$, $u(t)$ is a weather variable based input vector and $w(t)$ is a vector of random white noise input. Matrices A, B and the vector c are assumed to be constant. Some examples of state space method can be found in reference [8]. According to the reference [9], a potential advantage of state space method over ARIMA methods is the possibility to use a priori information in parameter estimation via Bayesian techniques. It is also pointed out that the advantages are not very clear and more experimental comparisons are needed.

4.3　Artificial Intelligence Methods

Artificial intelligence (AI) methods, such as expert system (ES), artificial neural network (ANN), genetic algorithm (GA), evolutionary computation (EC), fuzzy logic, etc. have been emerged in recent years in power systems applications as effective tools. In a practical power system, it is very important to have the human knowledge and experiences over a period of time due to various uncertainty, load variations, topology changes, etc.

This section presents the overview of AI methods (ES, ANN, fuzzy systems, EC, ant colony search, Tabu search, etc.) for power systems problems.

4.3.1 *Expert Systems*

AI techniques that achieve expert-level competence in solving the problems by bringing knowledge about specific tasks are called knowledge-based or expert systems (ES). ES was first proposed by Feigenbaum et al. in the early 1970s [10], [11]. ES is a knowledge-based or rule based system, which uses the knowledge and interface procedure to solve problems that are difficult enough to be solved by human expertise. Main advantages of ES [12] are:

(a) It is permanent and consistent,

(b) It can be easily transferred or reproduced and

(c) It can be easily documented.

Main disadvantage of ES is that it suffers from a knowledge bottleneck by having inability to learn or adapt to new situations. The knowledge engineering techniques started with simple rule based technique and extended to more advanced techniques such as object-oriented design, qualitative reasoning, verification and validation methods, natural languages, and multi-agent systems. For the past several years, a great deal of ES applications has been developed to prepare plans, analyze, manage, control and operate various aspects of power generation, transmission and distributions systems. Surveys of ES applications in power system are presented in reference [13]-[16]. Lu et al. [17], [18] have done considerable work on the applications of ES in power systems. Expert system has also been applied in recent years for load, bid and price forecasting.

4.3.2 *Artificial Neural Networks*

An artificial neural network (ANN) is an information-processing paradigm that is inspired by the way biological nervous systems, such as the brain, process information. The key element of this paradigm is the novel structure of the information processing system. It is composed of a large number of highly interconnected processing elements (neurons) working in unison to solve specific problems. ANNs, like people, learn by example. The starting point of ANN was the training algorithm proposed by Hebb in 1949, which demonstrated how a network of neurons could exhibit learning behaviour [18]. ANNs are mainly categorized by their architecture (number of layers), topology (connectivity pattern, feed forward or recurrent, etc.) and learning regime.

The main advantages of ANN [20]-[22] are as follows:

- It is fast,

- It possesses learning ability,

- It adapts to the data,

- It is robust and

- It is appropriate for non-linear modelling.

Though, the neural network training is generally computationally expensive, it takes negligible time to evaluate correct outcome once the network has been trained. Despite the advantages, some disadvantages of the ANN are: (i) large dimensionality, (ii) selection of the optimum configuration, (iii) choice of training methodology, (iv) the 'black-box' representation of ANN – they lack explanation capabilities and (v) the fact that results are always generated even if the input data are unreasonable. Vankayala et al. [23] have presented a bibliographical survey of neural network and their applications to power systems. Artificial neural networks are most promising methods for many power system problems. Various methods of ANN have been used for load, bid and price forecasting.

4.3.3 *Fuzzy Logic*

Fuzzy logic was developed by Zadeh [24] in 1964 to address uncertainty and imprecision, which widely exist in the engineering problems and it was first introduced in 1979 for solving power system problems. Fuzzy set theory can be considered as a generalization of the classical set theory. In classical set theory, an element of the universe either belongs to or does not belong to the set. Thus, the degree of association of an element is crisp. In a fuzzy set theory, the association of an element can be continuously varying. Mathematically, a fuzzy set is a mapping (known as membership function) from the universe of discourse to the closed interval [0, 1]. The membership function is usually designed by taking into consideration the requirements and constraints of the problem. Fuzzy logic implements human experiences and preferences via membership functions and fuzzy rules.

Due to the use of fuzzy variables, a non-expert operator can also be made to understand the system. In this way, fuzzy logic can be used as a general methodology to incorporate knowledge, heuristics or theory into controllers and decision makers. The advantages of fuzzy theory are as follows:

- It more accurately represents the operational constraints of power systems and
- Fuzzified constraints are softer than the traditional constraints [25], [26].

A detailed introduction to fuzzy logic and their applications in power systems has been presented in reference [27], [28]. Momoh et al. [29] have presented the overview and literature survey of fuzzy set theory application in power systems. A recent survey presented in reference [30] shows that fuzzy set theory has been applied mainly in voltage and reactive power control, load and price forecasting, fault diagnosis, power system protection/ relaying, stability and power system control, etc.

4.3.4 *Evolutionary Computation*

Natural evolution is a hypothetical population-based optimization process. Evolutionary computation (EC) is based on the Darwin's principle of 'survival of the fittest strategy'. An evolutionary algorithm begins by initializing a population of solutions to a problem [31]. New solutions are then created by randomly varying those of the initial population. All solutions are measured with respect to how well they address the task. Finally, a selection criterion is applied to weed out those solutions which are below standard. The process is iterated using the selected set of solutions until a specific criterion is met. The advantages of EC are adaptability to change and ability to generate good enough solutions but it needs to be understood in relation to computing requirements and convergence properties. EC can be subdivided into GA, evolution strategies, evolutionary programming (EP), genetic programming, classified systems, simulated annealing (SA), etc. The first work in the field of evolutionary computation (EC) was reported by Fraser in 1957 [32] to study the aspects of genetic system using a computer. After some time, a number of evolutionary inspired optimization techniques were developed, i.e. by Friedman in 1959 [33], Blendsoe in 1961 [34] and Bremermann in 1962 [35]. EC was presented by Fogel et al. in 1966 [36].

4.3.5 Genetic Algorithm

Genetic algorithm (GA) is a search algorithm based on the conjecture of natural selection and genetics. The features of genetic algorithm are different from other search techniques in several aspects. First, the algorithm is a multi-path that searches many peaks in parallel and thus reducing the possibility of local minimum trapping. Secondly, GA works with a coding of parameters instead of the parameters themselves. The coding of parameters will help the genetic operator to evolve the current state into the next state with minimum computations. Thirdly, GA evaluates the fitness of each string to guide its search instead of the optimization function. The genetic algorithm only needs to evaluate objective function (fitness) to guide its search. Hence, there is no need for computation of derivatives or other auxiliary functions. Finally, GA explores the search space where the probability of finding improved performance is high. It is distinguished from conventional optimization techniques by the use of concepts of population genetics to guide the optimization search. Instead of point-to-point search, GA searches from population to population.

The advantages of GA over traditional techniques are as follows:
- It needs only rough information of the objective function and puts no restriction

such as differentiability and convexity on the objective function.

- The method works with a set of solutions from one generation to the next and not a single solution, thus making it less likely to converge on local minima.
- The solutions developed are randomly based on the probability rate of the genetic operators such as mutation and crossover, as the initial solutions would not dictate the search direction of GA.

Major disadvantage of GA method is that it requires tremendously high computational time in the case of large variables and constraints. The treatment of equality constraints is also not well established in GA. Alander [37] has presented a bibliography of genetic algorithm used in power systems.

4.3.6 Evolution Strategies and Evolutionary Programming

Evolution strategy (ES) employs real-coded variables and in its original form, it relied on mutation as the search operator and a population size of one. Since then, it has evolved to share many features with GA. The major similarity between these two types of algorithms is that they both maintain populations of potential solutions and use a selection mechanism for choosing the best individuals from the population. The main differences are:

- ES operates directly on floating point vectors while classical GA operates on binary strings,
- GA relies mainly on recombination to explore the search space while ES uses mutation as the dominant operator and
- ES is an abstraction of evolution at individual behaviour level, stressing the behavioural link between an individual and its offspring, while GA maintains the genetic link.
- Evolutionary programming (EP), which is a stochastic optimization strategy similar to GA, places emphasis on the behavioural linkage between parents and their offspring, rather than seeking to emulate specific genetic operators as observed in

nature. EP is similar to evolutionary strategies, although the two approaches were developed independently. Like ES and GA, EP is a useful method of optimization when other techniques such as gradient descent or direct analytical discovery are not possible. Combinatorial and real-valued function optimization, in which the optimization surface or fitness landscape is "rugged" and possesses many locally optimal solutions, is well suited for evolutionary programming.

4.3.7 *Simulated Annealing*

Based on the annealing process in the statistical mechanics, the simulated annealing (SA) was introduced for solving complicated combinatorial optimization problems. In a large combinatorial optimization problem, an appropriate perturbation mechanism, cost function, solution space and cooling schedule are required in order to find an optimal solution with simulated annealing. SA is effective in network reconfiguration problems for large-scale distribution systems and its search capability becomes more significant as the system size increases. Moreover, the cost function with a smoothing strategy enables the simulated annealing to escape more easily from local minima and reach rapidly to the vicinity of an optimal solution.

The advantages of SA are its general applicability to deal with arbitrary systems and cost functions, its ability to refine optimal solution and its simplicity of implementation even for complex problems. The major drawback of SA is repeated annealing. This method cannot tell whether it has found optimal solution or not. Some other method (e.g. branch and bound) is required to do this. SA has been used in various power system applications like transmission expansion planning [38]-[40], unit commitment [41]-[43], maintenance scheduling [44] etc.

4.3.8 *Ant Colony Search*

Dorigo introduced the ant colony search (ACS) system, for the first time, in 1992 [45]. ACS techniques take inspiration from the behavior of real ant colonies and are used to solve function or combinational problems. ACS algorithms to some extent mimic the behavior of real ants. The main characteristics of ACS are positive feedback for recovery of good solutions, distributed computation, which avoids premature convergence and the use of a constructive heuristic to find acceptable solutions in the early stages of the search process. The main drawback of the ACS technique is poor computational features. ACS technique has been mainly used in finding the shortest route for transmission network [46].

4.3.9 *Tabu Search*

Tabu search (TS) is basically a gradient-descent search with memory. The memory preserves a number of previously visited states along with a number of states that might be considered unwanted. This information is stored in a Tabu list. The definition of a state, the area around it and the length of the Tabu list are critical design parameters. In addition to these Tabu parameters, two extra parameters are often used such as aspiration and diversification. Aspiration is used when all the neighbouring states of the current state are also included in the Tabu list. In that case, the Tabu obstacle is overridden by selecting a new state. Diversification adds randomness to this otherwise deterministic search. If the Tabu search is not converging, the search is reset randomly.

TS is an iterative improvement procedure that starts from some initial solution and attempts to determine a better solution in the manner of a 'greatest descent neighbourhood' search algorithm. Basic components of TS are the moves, tabu list and aspiration level. TS is a meta-huristic search to solve global optimization problem, based on multi-level memory management and response exploration [47]-[50]. TS has been used in various power system application like transmission planning [51], optimal capacitor placement [52]-[54], unit commitment [55], hydrothermal scheduling [56], fault diagnosis/alarm processing [57]- [58], reactive power planning [59] etc.

4.3.10 *Particle Swarm Optimization*

Particle swarm optimization (PSO) is an exciting new methodology in evolutionary computation that is somewhat similar to a genetic algorithm in that the system is initialized with a population of random solutions. Unlike other algorithms, however, each potential solution (called a particle) is also assigned a randomized velocity and then flown through the problem hyperspace. Particle swarm optimization has been found to be extremely effective in solving a wide range of engineering problems. It is very simple to implement (the algorithm comprises two lines of computer code) and solves problems very quickly.

4.4 Artificial Neural Network (ANN) Models

Artificial neural network are most widely used in forecasting electricity market parameters. Various ANN models used worldwide by researchers are listed and discussed below.

4.4.1 *Feed-Forward Neural Network*

The feed-forward neural network is the first and simplest type of artificial neural network. In this network, the information move in only one direction (forward) from the input nodes, through the hidden nodes (if any), to the output nodes. There are no cycles or loops in the network. There are different variants of this network like single layer perceptron, multi-layer perceptron and ADALINE (adaptive linear element).

4.4.2 Radial Basis Function (RBF) Network

Radial basis functions are powerful techniques for interpolation in multi-dimensional space. A RBF is a function which has built into a distance criterion with respect to a center. Radial basis functions have been applied in the area of neural networks where they may be used as a replacement for the sigmoidal hidden layer transfer characteristic in multi-layer perceptron (MLP). RBF networks (RBFNs) have two layers of processing. In the first, input is mapped onto each RBF in the 'hidden' layer. The RBF chosen is usually a Gaussian function. In regression problems, the output layer is then a linear combination of hidden layer values representing mean predicted output. The interpretation of this output layer value is the same as a regression model in statistics. In classification problems, the output layer is typically a sigmoid function of a linear combination of hidden layer values, representing a posterior probability. Performance in both cases is often improved by shrinkage techniques, known as ridge regression in classical statistics and known to correspond to a prior belief in small parameter values (and therefore smooth output functions) in a Bayesian framework.

RBF networks have the advantage of not suffering from local minima in the same way as multi-layer perceptron model. This is because the only parameters that are adjusted in the learning process are the linear mapping from hidden layer to output layer. Linearity ensures that the error surface is quadratic and therefore, a single optimal is easily found. In regression problems, this can be found in one matrix operation. In classification problems, the fixed non-linearity introduced by the sigmoid output function is the most efficiently dealt with using iteratively re-weighted least squares.

RBF networks have the disadvantage of requiring good coverage of the input space by radial basis functions. RBF centers are determined with reference to the distribution of the input data, but without reference to the prediction task. As a result, representational resources may be wasted on areas of the input space that are irrelevant to the learning task. A common solution is to associate each data point with its own centre, although this can make the linear system to be solved in the final layer and requires shrinkage techniques to avoid over-fitting.

Associating each input datum with an RBF leads naturally to kernel methods such as support vector machines (SVMs) and Gaussian processes (the RBF is the kernel function). All three approaches use a non-linear kernel function to project the input data into a space where the learning problem can be solved using a linear model. Like Gaussian processes, and unlike SVMs, RBF networks are typically trained in a maximum likelihood framework by maximizing the probability (minimizing the error) of the data under the model. SVMs take a different approach to avoiding over-fitting by maximizing a margin. RBF networks are out-performed in most classification applications by SVMs. In regression applications, these can be competitive when the dimensionality of the input space is relatively small.

4.4.3 *Kohonen's Self-Organizing Network*

The self-organizing map (SOM) invented by Teuvo Kohonen is a form of unsupervised learning. A set of artificial neurons learn to map points in an input space to coordinates in an output space. The input space can have different dimensions and topology from the output space and the SOM will attempt to preserve these.

4.4.4 *Recurrent Network*

Contrary to feed-forward networks, recurrent neural networks (RNNs) are models with bi-directional data flow. While a feed-forward network propagates data linearly from input to output, RNNs also propagate data from later processing stages to earlier stages. A simple recurrent network (SRN) is a variation on the multi-layer perceptron network (MLPN), sometimes called an Elman network due to its inventor, Jeff Elman. A three-layer network is used, with the addition of a set of context units in the input layer. There are connections from the middle (hidden) layer to these context units fixed with a weight of one. At each time step, the input is propagated in a standard feed-forward fashion and then a learning rule (usually back-propagation) is applied. The fixed back connections result in the context units always maintaining a copy of the previous values of the hidden units. Thus, the network can maintain a sort of state, allowing it to perform such tasks as sequence-prediction that are beyond the power of a standard multi-layer perceptron.

In a fully recurrent network, every neuron receives inputs from every other neuron in the network. These networks are not arranged in layers. Usually only a subset of the neurons receive external inputs in addition to the inputs from all the other neurons and another disjunct subset of neurons report their output externally as well as sends it to all the neurons. These distinctive inputs and outputs perform the function of the input and output layers of a feed-forward or simple recurrent network and also join all the other neurons in the recurrent processing.

4.4.5 Hopfield Network

The Hopfield network is a recurrent neural network in which all connections are symmetric. Invented by John Hopfield in 1982, this network guarantees that its dynamics will converge. If the connections are trained using Hebbian learning, then the Hopfield network can perform as robust content-addressable (or associative) memory, resistant to connection alteration.

4.4.6 Stochastic Neural Networks

A stochastic neural network differs from a typical neural network because it introduces random variations into the network. In a probabilistic view of neural networks, such random variations can be viewed as a form of statistical sampling, such as Monte Carlo simulation.

4.4.7 Boltzmann Machine

The Boltzmann machine can be thought of a noisy Hopfield network. Invented by Geoff

Hinton and Terry Sejnowski in 1985, the Boltzmann machine is important because it is one of the first neural networks to demonstrate learning of latent variables (hidden units). Boltzmann machine learning was at first slow to simulate, but the contrastive divergence algorithm of Geoff Hinton allows models such as Boltzmann machines and products of experts to be trained much faster.

4.4.8   Committee of Machines

A committee of machines (CoM) is a collection of different neural networks that together vote on a given example. This, generally, gives a much better result compared to other neural network models. Because neural networks suffer from local minima, starting with the same architecture and training but using different initial random weights, often gives vastly different networks.

The CoM is similar to the general machine learning method, except that the necessary variety of machines in the committee is obtained by training from different random starting weights rather than training on different randomly selected subsets of the training data.

4.4.9   Associative Neural Network (ASNN)

The ASNN, which is an extension of the committee of machines, goes beyond a simple/weighted average of different models. ASNN represents a combination of an ensemble of feed-forward neural networks and the k-nearest neighbor (kNN) technique. It uses the correlation between ensemble responses as a measure of distance amid the analyzed cases for the kNN. This corrects the bias of the neural network ensemble. An associative neural network has a memory that can coincide with the training set. If new data become available, the network instantly improves its predictive ability and provides data approximation (self-learn the data) without a need to retrain the ensemble. Another important feature of ASNN is the possibility to interpret neural network results by analysis of correlations between data cases in the space of models.

4.4.10 *Cascading Neural Networks*

Cascade-correlation is a supervised learning algorithm developed by Scott Fahlman and Christian Lebiere. Instead of just adjusting the weights in a network of fixed topology, cascade-correlation begins with a minimal network, then automatically trains and adds new hidden units one by one, creating a multi-layer structure [60]. Once a new hidden unit has been added to the network, its input-side weights are frozen. This unit then becomes a permanent feature-detector in the network, available for producing outputs or for creating other, more complex feature detectors. The cascade-correlation architecture has several advantages over existing algorithms such as:

- it learns very quickly i.e. the network determines its own size and topology,

- it retains the structures,

- it can be built even if the training set changes and

- it requires no back-propagation of error signals through the connections of the network.

4.4.11 *Neuro-Fuzzy Networks*

A neuro-fuzzy network (NFN) is a fuzzy inference system in the body of an artificial neural network. Depending on the fuzzy inference system (FIS) type, there are several layers that simulate the processes involved in a fuzzy inference like fuzzification, inference, aggregation and defuzzification. Embedding an FIS in a general structure of an ANN has the benefit of using available ANN training methods to find the parameters of a fuzzy system.

4.5 Summary

Conventional and modern AI techniques have been discussed for load forecasting, price forecasting, bid forecasting and spinning reserve (SR) forecasting in power system. Based on the available literature, it can be stated that the conventional methods are not very suitable to forecast modern day electricity market variables in view of the uncertainty and variations. AI relies heavily on good problem description and extensive domain knowledge. ES, which is a knowledge-based system, suffers from a knowledge bottleneck by having an inability to learn or to adapt to new situations. Knowledge-based system can enhance the capabilities of a power system, whereas ANN can acquire knowledge through adaptive training and generalization. ANN will be the most useful AI technique for load and price forecasting. Various ANN models, used by engineers and researchers world-wide have been discussed. A feed forward multi-layer perceptron (MLP) model with back propagation (BP) training algorithms (gradient descent or with conjugate gradient algorithm) has been applied for electricity load, price, bid and SR forecasting. Cascaded architectures of multiple ANNs and committee machine replace the single neural network for complex nonlinear mapping functions. Radial basis function neural network (RBFNN) and recurrent neural network (RNN) have also been proposed for forecasting due to several advantages. A new fuzzy-neural network (FNN) technique with higher learning capability has been proposed to forecast market parameters. Fuzzy theory with its realistic description of power system problems and ANN with its promise of adaptive training and generalization, deserves the scope for further study. The hybrid AI techniques seem to be getting most of the attention. The application of hybrid systems in power system problems is a novel development, which represents a definite future trend in providing solutions to the power system problems.

References

[1]   Mohd. Asif Iqbal, "Analysis and Comparison of Lambda Iteration, Genetic Algorithm and Particle Swarm Optimization to Solve Economic Load Dispatch Problem", International Journal of Software and Web Sciences (IJSWS), pp 60-64

[2]   M. M. Tripathi, K. G. Upadhyay, S. N. Singh, "Short-Term Load Forecasting using Generalized Regression and Probabilistic Neural Networks in the Electricity Market", The Electricity, Volume 21, Issue 9, November 2008, pp 24-34

[3]   Pauli Murto, "Neural network models for short-term load   forecasting",   Masters   thesis,

Department of Engineering Physics and Mathematics, Helisinki University of technology, Helsinki, January 5, 1998.

[4] M. Räsänen and J. Ruusunen, "Verkoston tilan seurantamittauksilla jakuormitusmalleilla", Research Report B17, Systems Analysis Laboratory, Helsinki University of Technology.

[5] A. D. Papalexopoulos and T. C. Hesterberg, "A regression-based approach to short-term system load forecasting", IEEE Trans. on Power Systems, Vol. 5, No. 4, November 1990, pp. 1535-1547.

[6] M. T. Hagan and S. M. Behr, "The time series approach to short term load forecasting", IEEE Trans. on Power Systems, Vol. PWRS-2, No. 3, August 1987, pp. 785-791.

[7] R. S. Pindyck and D. L. Rubinfeld, "Econometric models and economic forecasts", McGraw-Hill, Singapore.

[8] R. Campo and P. Ruiz, "Adaptive weather-sensitive short-term load forecast", IEEE Trans. on Power Systems, Vol. PWRS-2, No. 3, August 1987, pp. 592-600.

[9] G. Gross and F. D. Galiana, "Short-term forecasting", Proceedings of the IEEE, Vol. 75, No. 12, 1987, pp. 1558-73.

[10] B. G. Bucannan and E. A. Feigenbaum, "Dendral and Metadendral: Their Applications Dimension", Journal of Artificial Intelligence, November 1978, pp. 5-24.

[11] T. J. Dillon and M. A. Laughton, Expert System Application in Power Systems", Prentice Hall, London, 1990.

[12] K. Wardwick, A. Ekwue and R. Aggarwal, "Artificial Intelligence Techniques in Power Systems", IEE Power Engineering Series, 22, London, UK, 1997.

[13] Z. Z. Zhang, G. S. Hope and O. P. Malik, "Expert Systems in Electric Power Systems - A Bibliographic Survey" IEEE Trans. Power Systems, Vol. 4, No. 4, 1989, pp. 1355-1362.

[14] N. J. Balu, R. A. Adapa, G. Cauley, M. Lauby and D. J. Maratukulam, "Review of Expert Systems Planning and Operations", Proceedings of IEEE, Vol. 80, No. 5, 1992, pp. 727-731.

[15] CIGRE Task Force 38-06-02, "Survey on Expert System in Alarm Handling", Electra, Vol. 139, 1991, pp. 133-151.

[16] A. J. Germond and D. Niebur, "Survey of Knowledge-Based Systems in Power Systems: Europe", Proceedings of IEEE, Vol. 80, No. 5, 1992, pp. 732-744.

[17] C. C. Liu, T. K. Ma, K. L. Liou and M. S. Tsai, "Practical Use of Expert Systems in Power Systems", International Journal on Engineering Intelligent Systems, Vol. 2, No. 1, 1994, pp. 11-22.

[18] C. C. Liu, "Practical Use of Expert Systems in Planning and Operation of Power Systems", Electra, Vol. 141, February 1993, pp. 31-67.

[19]   D. Niebur and T. S. Dillon, "Neural Network Applications in Power Systems", CRL Publishing Ltd, U.K., 1996.

[20]   R. Aggarwal and Y. H. Song, "Artificial Neural Networks in Power Systems-Part 1: General Introduction to Neural Computing", IEE Power Engineering Journal, Vol. 11, No. 3, 1997, pp. 129-134.

[21]   R. Aggarwal and Y. H. Song, "Artificial Neural Networks in Power Systems-Part 2: Types of Artificial Neural Networks", IEE Power Engineering Journal, Vol. 12, No. 1, 1998, pp. 41-47.

[22]   R. Aggarwal and Y.H. Song, "Artificial Neural Networks in Power Systems-Part 3: Examples of Applications in Power Systems', IEE Power Engineering Journal, Vol. 12, No. 6, 1998, pp. 279-287.

[23]   V. S. S. Vankayala and N. D. Rao, "Artificial Neural Networks and Their Applications to Power Systems - A Bibliographical Survey", Electric Power Systems Research, Vol. 28, No. 1, 1993, pp. 67-79.

[24]   L. A. Zadeh, "Fuzzy Sets, Information and Control", Vol. 8, 1965, pp. 338-353.

[25]   S. K. Pal and D. P. Mandal, "Fuzzy Logic and Approximate Reasoning: An Overview", Journal of Institution of Electronics and Telecommunication Engineers, Paper No. 186-C, 1991, pp. 548-559.

[26]   Y. H. Song and A. T. Johns, "Applications of Fuzzy Logic in Power Systems-Part I: General Introduction to Fuzzy Logic", IEE Power Engineering Journal, Vol. 11, No. 5, 1997, pp. 219-222.

[27]   Y. H. Song and A. T. Johns, "Applications of Fuzzy Logic in Power Systems-Part II: Comparison and Integration with Expert Systems Neural Networks and Genetic Algorithms", IEE Power Engineering Journal, Vol. 12, No. 4, 1998, pp. 185-190.

[28]   Y. H. Song and A. T. Johns, "Applications of Fuzzy Logic in Power Systems-Part III: Example Applications", IEE Power Engineering Journal, Vol. 13, No. 2, 1999, pp. 97-103.

[29]   J. A. Momoh, X. W. Ma and K. Tomsovic, "Overview and Literature Survey of Fuzzy Set Theory in Power Systems" IEEE Trans. Power Systems, Vol. 10, No. 3, 1995, pp. 1676-1690.

[30]   R. C. Bansal, "Bibliography on the Fuzzy Set Theory Applications to Power Systems (1994-2001), IEEE Trans. Power Systems, Vol. 18, No. 4, 2003, pp. 1291-1299.

[31]   D. B. Fogel, "What is Evolutionary Computation?", IEEE Spectrum, Vol. 37, No. 2, 2000, pp. 26-32.

[32]   A. S. Fraser, "Simulation of Genetic Systems by Automatic Digital Computers" Australian Journal of Biological Sciences, Vol. 10, 1957,   pp. 484-491.

[33] G. D. Friedman, "Digital Simulation of an Evolutionary Process" General   Systems Yearbook, 4, 1959.

[34]   W. W. Blendsoe, "The Use of Biological Concepts in the Analytical Study of Systems", Paper presented at the ORSA-TIMS National   Meeting, San Francisco, CA, Nov. 1961.

[35]   H. J. Bremermann, "Optimization Through Evolution and   Recombination" In M. C. Yovits, G. T. Jacobi, and G. D. Goldstein (editors), Self Organization Systems, Spartan Books, Wasington, D. C., 1962, pp. 93-106.

[36]   L. J. Fogel, A. J. Owens, and M. J. Walsh, "Artificial Intelligence Through Simulated Evolution", John Wiley, New York, 1966.

[37]   J. T. Alander, "An Indexed Bibliography of Genetic Algorithm in Power Engineering' Report Series 94-1, Feb. 1996.

[38]   R. A. Gallego, A. B. Alves, A. Monticelli, and R. Romero, "Parallel Simulated Annealing Applied to Long Term Transmission Network   Expansion Planning" IEEE Trans. Power Systems, Vol. 12, No. 1, 1997, pp. 181-188.

[39]   R. Romero, R. A. Gallego and A. Monticelli, "Transmission System Expansion Planning by Simulated Annealing", IEEE Trans. Power Systems, Vol. 11, No. 1, 1996, pp. 364-369.

[40]   U. D. Annakkage, T. Numnonda and N. C. Pahalawaththa, "Unit Commitment by Parallel Simulated Annealing", IEE Proc. Generation, Transmission and Distribution, Vol. 142, No. 6, 1995, pp. 595-600.

[41]   A. H. Mentaway, Y. L. Abdel-Magid and S. Z. Selim, "A Simulated Annealing Algorithm for Unit Commitment", IEEE Trans. Power Systems, Vol. 13, No. 1, 1998, pp. 197-204.

[42]   H. T. Yang, P. C. Yang and C. L. Huang, "A Parallel Algorithm   Approach to Solving the Unit Commitment Problem: Implementation on the Transputer Networks", IEEE Trans. Power Systems, Vol. 12, No. 2, 1997, pp. 661-668.

[43]   T. Satoh and K. Nara, "Maintenance Scheduling by Using Simulated Annealing Method" IEEE Trans. Power Systems, Vol. 6, No. 2, 1991, pp. 850-857.

[44]   Y. H. Song and M. R. Irving, "Optimization Methods for Electric Power Systems: Part 2 Heuristic Optimization Methods", IEE Power   Engineering Journal, Vol. 15, No. 3, 2001, pp. 151-160.

[45]   M. Dorigo, V. Maniezzo and A. Colorni, "The Ant System: Optimization by a   Colony of Co-Operating Agents", IEEE Trans. Systems, Man and Cybernetics, Vol. 26, No. 1, 1996, pp. 29-41.

[46]   Y. H. Song and C. S. Chou, "Application of Ant Colony Search   Algorithms   in   Power

System Optimization", IEEE Power   Engineering Review, Vol. 18, No. 12, 1998, pp. 63- 64.

[47] F. Glover and M. Laguna, "Tabu Search", Kluwer Academic Publishers, Boston, USA, 1997.

[48] F. Glover, M. Laguna, E. Tallard and D.de Werra, "Tabu Search", Science Publishers, Basel, Switzerland, 1993.

[49] F. Glover, "Tabu Search – Part 1", ORSA Journal on Computing, Vol. 1, No. 3, 1989, pp. 190- 206.

[50] F. Glover, "Tabu Search – Part 2", ORSA Journal on Computing, Vol. 2, No. 1, 1990, pp. 4-32.

[51] F. Wen and C. S. Chang, "Transmission Network Optimal Planning using Tabu   Search Method", Electric Power System Research, Vol. 42, No. 2, 1997, pp. 153-163.

[52] Y. C. Huang, H. T. Yang and C. L. Haung, "Solving the Capacitor Placement Problem in a Radial Distribution System using Tabu Search Approach", IEEE Trans. Power Systems, Vol. 11, No. 4,   1996, pp. 1868-1873.

[53] H. T. Yang, Y. C. Huang and C. L. Haung, "Solution to Capacitor Placement Problem in a Radial Distribution System Using Tabu Search Method", Proc. of International Conference on Energy Management and Power Delivery, Singapore, 1995, pp. 388- 393.

[54] D. Gan, Y. Hayashi and K. Nara, "Multi-Level Reactive Resource Planning by   Tabu Search", Proc. of IEE of Japan Power and Energy, Nagoya, Japan, 1995, pp. 137-142.

[55]   H. Mori and T. Usami, "Unit Commitment using Tabu Search with Restricted Neighborhood", Proc. of International Conference on   Intelligent Systems Applications to Power Systems, Orlando, USA, 1996, pp. 422-427.

[56] X. Bai and S. Shahidehpour, "Hydro-Thermal Scheduling by Tabu Search and by Decomposition Method", IEEE Trans. Power   Systems, Vol. 11, No. 2, 1996,   pp. 968- 974.

[57] F. Wen and C. S. Chang, "A Tabu Search Approach to Fault Section Estimation   in   Power Systems", Electric Power System Research, Vol. 40, No. 1, 1997, pp. 63-73.

[58] F. Wen and C. S. Chang, "Tabu Search Approach to Alarm Processing in Power Systems", IEE Proc. Generation, Transmission and Distribution, Vol. 144, No. 1, 1997, pp. 31-38.

[59] D. Gan, Z. Qu and H. Cai, "Large Scale VAR Optimization by Tabu Search", Electric Power System Research, Vol. 39, No. 3, 1995, pp. 195-204.

[60] http://e-spacio.uned.es:8080/fedora/get/tesisuned:CiencEcoEmp-Mnoguer/Documento.pdf

CHAPTER 5

Short-Term Load Forecasting in Electricity Markets

5.1 Introduction

Short-term load forecasting with a lead time of up to few days ahead in the electricity markets is performed for the economic and secure operation of electric power system [1], which allows electric utilities to optimize resources for better energy prices. The system load to be forecasted is a random, non-stationary process, composed of thousands of individual components. Therefore, the scope of possible approaches to the load forecasting is wide.

Some of the most popular methods are time-of-day method, regression method, stochastic time series methods, state-space method, expert systems, artificial neural networks, etc. Artificial neural network (ANN) is very successful and widely used in forecasting applications. There exists a variety of ANNs like back propagation, radial basis function network (RBFN), recurrent neural network (RNN), etc. A generalized regression neural network (GRNN), which has a radial basis layer and a special linear layer, is often used for function approximation. Probabilistic neural networks (PNN) can be used for classification problems.

The most important work in building an ANN load forecasting model is the selection of input variables. An issue in load forecasting modelling can be the interdependence between price and load. This will be reflected in pricing patterns of the market. Since the relationship between electricity price and load is complex and dynamic, it is important to know how different customers' price response characteristics and locations affect the load forecasting. The most important input variables, which affect the load forecasting, are weather temperature and price as these are having a strong correlation with load.

5.2 ANN Forecasting Models

Many different artificial neural networks are used for short-term load forecasting. The load forecasting will be explained with two methods in following sections.

5.2.1 *Generalized Regression Neural Network*

Generalized regression neural network (GRNN) is a new-type of neural network suggested by Donald F. Specht in 1991. The theoretical foundation [2] of GRNN is based on non-linear regression analysis. If joint probability density function of random vector x_0 and random vector y is $f(x_0, y)$, regression value of y to x_0 is $\hat{y}(x_0)$ defined as

$$\hat{y}(x_0) = \frac{\int_{-\infty}^{\infty} y f(x_0, y) \, dy}{\int_{-\infty}^{\infty} f(x_0, y) \, dy}$$

(5.1)

Using Parzen distribution, density function $f(x_0, y)$ can be obtained according to (5.1) by sample data collection $\{x_i, y_i\}_{i=1}^{n}$.

$$f(x_0, y) = \frac{1}{n(2\pi)^{\frac{p+1}{2}} w_1 w_2 \ldots w_p w_y} \sum_{i=1}^{n} e^{-d(x_0, x_i)} e^{-d(y, y_i)}$$

(5.2)

where, n is sample capability, p is dimension of x, w is spread factor of Gauss function, which is a smooth parameter and

$$d\left(x_0, x_i\right) = \sum_{j=1}^{p} \left[\left(x_{oj} - x_{ij}\right) \Big/ w_{ij} \right]^2 , \quad d\left(y, y_i\right) = \left[y - y_i\right]^2 \tag{5.3}$$

Putting (5.2) into (5.1), $\hat{y}\left(x_0\right)$ can be obtained as

$$\hat{y}\left(x_o\right) = \frac{\displaystyle\sum_{i=1}^{n} \left(e^{-d\left(x_0, x_i\right)} \int_{-\infty}^{+\infty} y\, e^{-d\left(y, y_i\right)} dy \right)}{\displaystyle\sum_{i=1}^{n} \left(e^{-d\left(x_0, x_i\right)} \int_{-\infty}^{+\infty} e^{-d\left(y, y_i\right)} dy \right)} \tag{5.4}$$

Since $\displaystyle\int_{-\infty}^{\infty} z e^{-z^2} dz = 0$, $\hat{y}\left(x_0\right)$ can be simplified as

$$\hat{y}\left(x_0\right) = \frac{\displaystyle\sum_{i=1}^{n} y_i\, e^{-d\left(x_0, x_i\right)}}{\displaystyle\sum_{i=1}^{n} e^{-d\left(x_0, x_i\right)}} \tag{5.5}$$

The architecture for the GRNN is shown in fig. 5.1 and it is similar to the radial basis network, but has a slightly different second layer. The first layer has as many neurons as input/ target vectors. Each neuron's weighted input is the distance between the input vector and its weight vector. Each neuron's net input is the product of its weighted input with its bias.

The output of each neuron is its net input passed through radial basis layer. The second layer has also as many neurons as input/target vectors. A larger spread (distance) leads to a large area around the input vector where layer 1 neurons will respond with significant outputs. Therefore, if spread is small, the radial basis function is very steep so that the neuron with the weight vector closest to the input will have a much larger output than other neurons. The network will tend to respond with the target vector associated with the nearest design input vector. As spread gets larger, the radial basis function's slope gets smoother and several neurons may respond to an input vector. The network then acts as a weighted average of target vectors whose designed input vectors are closest to the new input vectors.

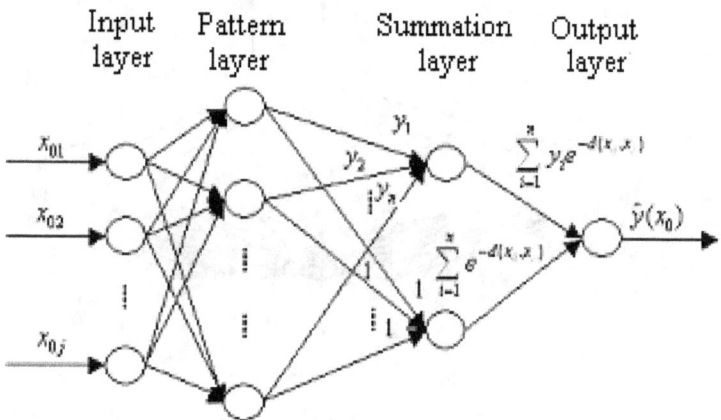

Fig. 5.1: General regression neural network (GRNN)

5.2.2 Probabilistic Neural Network

Probabilistic neural network is a kind of radial basis network suitable for classification problems [3]. Structure of a probabilistic neural network is shown in fig. 5.2. The probabilistic neural network (PNN) constitutes an alternative approach for class conditional density estimation. It is an RBF-like neural network adopted to provide output values corresponding to the class conditional densities. Since the network is RBF, the components (hidden units) are shared among classes and each class conditional density is evaluated using not only the corresponding class data points (as in the case of separate mixtures) but also all the available data points.

Probabilistic neural networks can be used for classification problems. When an input is presented, the first layer computes distances from the input vector to the training input vectors and produces a vector whose elements indicate how close the input is to a training input [4]. The second layer sums these contributions for each class of inputs to produce its net output vector of probabilities. Finally, a complete transfer function on the output of the second layer picks the maximum of these probabilities, and produces one for that class and a zero for the other classes.

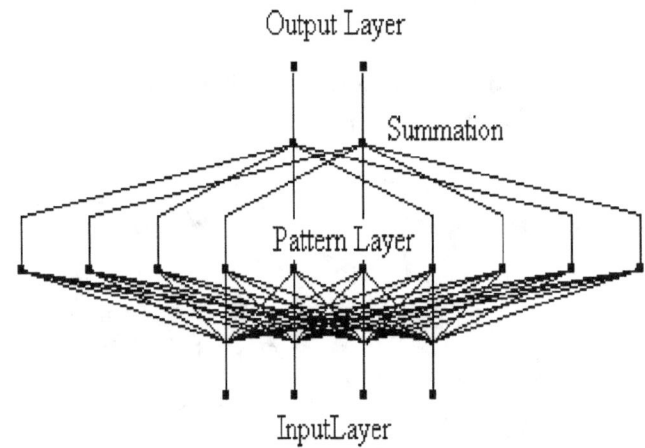

Fig. 5.2: Probabilistic neural network (PNN)

5.3 Input Variables to Artificial Neural Network

The selection of input variables for building an ANN load-forecasting model is very crucial. There is no general rule that can be followed in this process. It depends on engineering judgment and experience, and is carried out almost entirely by trial and error method [5]. However, some statistical analysis can be very helpful in determining variables that have significant influence on the system load. In general, three types of variables are used as inputs to the neural network: (a) hour and day indicators, (b) weather related inputs, (c) historical loads and pricing signal.

5.3.1 Hour and Day Indicators

Load varies throughout the day, which can be seen from fig. 5.3 below.

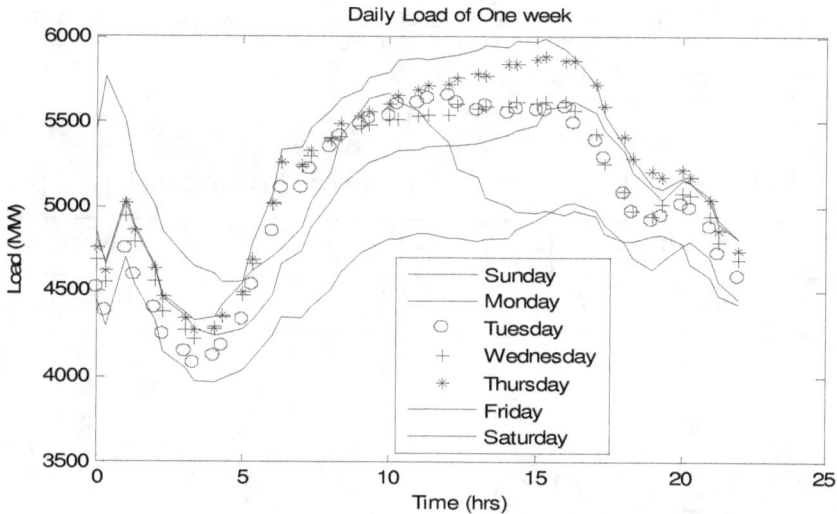

Fig. 5.3: Load curves during a typical day

There may be several peaks and valleys during a day and there is a significant difference in load magnitude. Therefore, an hour indicator $H(k)$ (where k changes from 1 to 24) is very helpful in short term load forecasting. Load also changes from day to day during a week. Fig. 5.3 also shows daily load for a typical winter week. It is clear from the figure that power consumption is different on different days. As a result, a day indicator $D(k)$ (where k changes from 1 to 7) is helpful in load forecasting.

5.3.2 Weather Variables

Temperature is the most important weather variable and only temperature is considered in this application. Other weather variables such as dew point, atmospheric pressure, wind velocity, cloud cover, etc. are neglected, as these have less effect.

5.3.3 Historical Load and Price

The price-load relationship is neither linear nor stationary in time but can be expected that price-load relationship is relatively stable over shorter periods of time. Volatile electricity prices in power markets are new phenomenon that needs to be examined. It is clear from fig. 5.4 and table 5.1 that there is a strong correlation between load and price. Thus, price is a major deciding factor for load forecasting.

Table 5.1: Correlation factor between load and historical price

Spain	PJM	NEMMCO
0.8993	0.9397	0.8517

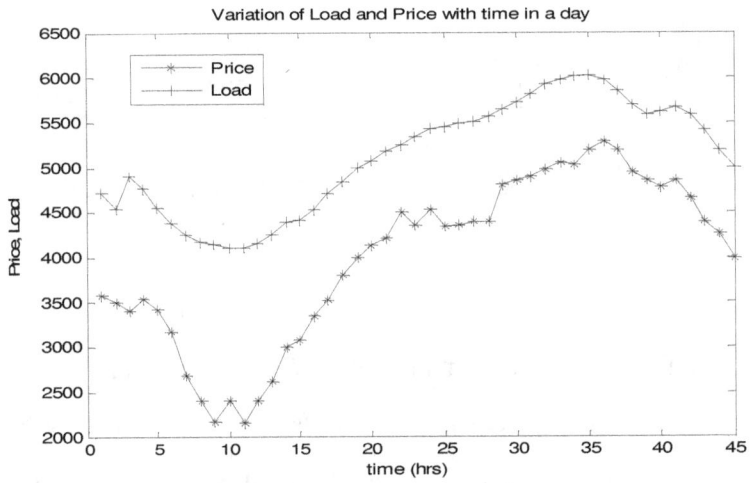

Fig 5.4: Load and price changes during a typical day

5.4 ANN Training Data and Algorithm

There are many electricity markets operating in the world. The NEMMCO electricity market is covering Victoria, New South Wales, Queensland, South Australia and the Australian Capital Territory. In the USA, PJM Interconnection plays a vital role in the U.S. electric system and responsible for the operation of the largest centrally dispatched electric system in North America. The PJM market is coordinated by an independent system operator (ISO). The ISO ensures a secured, economical and efficient operation as well as determining all locational marginal prices (LMP) according to voluntary bids from the market participants.

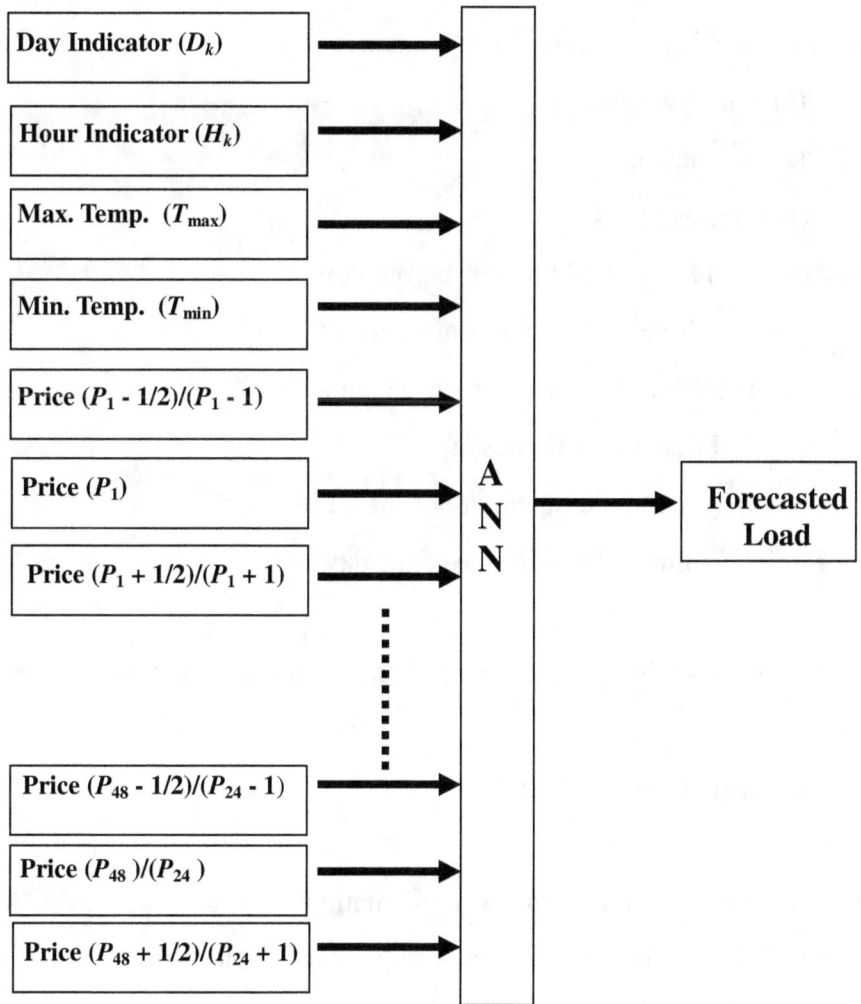

Fig. 5.5: The ANN model used in load forecasting

Publicly available historical data from the web sites of electricity markets can be used for to forecast the load. The data are divided into several windows where most of these are used for training and the remaining data are used for testing the ANN. More precisely, for each month, the first, second and fourth weeks are used for training, while the third week is left for testing of the ANN. Training was done for all the data windows at the same time i.e. the same ANN trained is to be used at any time during the year. All inputs and outputs are normalized before training. Cross correlation between load and price is found and only those price inputs are considered which are the best correlated.

The inputs to the ANN as shown in fig. 5.5 are:

- $H(k)$ hour indicator
- $D(k)$ day indicator
- $P(k)$ Price at hour k
- $P(k - 1/2)$ Price at 30 minutes before hour k
- $P(k + 1/2)$ Price at 30 minutes after hour k
- $P(k - 1)$ Price at 60 minutes before hour k
- $P(k + 1)$ Price at 60 minutes after hour k
- $T(\max)$ Maximum temperature of the day
- $T(\min)$ Minimum temperature of the day

Training of the ANN is performed by presenting the set of input/output data.

5.5   Simulation and Results

Simulation is carried out using any simulation software such as MATLAB. First, the developed ANN models are trained using set of input/ output data. Due to its special features as explained in the previous sections, the algorithm resulted in a very fast training and the error is significantly reduced to very low value. Then, performance of the developed ANN models for forecasting load profile is tested using windows of data that are not included in the training set.

For more accurate evaluation of the ANN performance, the following absolute percentage error is used and defined as:

$$e = \frac{\text{Actual load - Forecasted load}}{\text{Actual Load}} * 100 \qquad (5.6)$$

An average of the absolute error over a period of time may be used for an overall evaluation and comparison with other techniques. The mean absolute percentage error is given in (5.7) where L_A is the actual load, L_F is the forecasted load, N is the number of hours and i is the hour index.

$$MAPE(\%) = \frac{1}{N} \sum_{i=1}^{N} \frac{\left| L_F^i - L_A^i \right|}{L_A^i} * 100 \qquad (5.7)$$

Some of the results of short-term forecasting in Victorian electricity market is shown here. Fig. 5.6 (a), (b), (c), (d), (e), (f) and (g) show actual and forecasted load of Victorian electricity market for each day of the week during 15-21 January 2006 with generalized regression neural network. The actual load curve of all the days of the week has a dip in the early morning, reaches its maximum during day time and don't vary much during that time. The forecasted load curves also follow the same pattern on all weekdays.

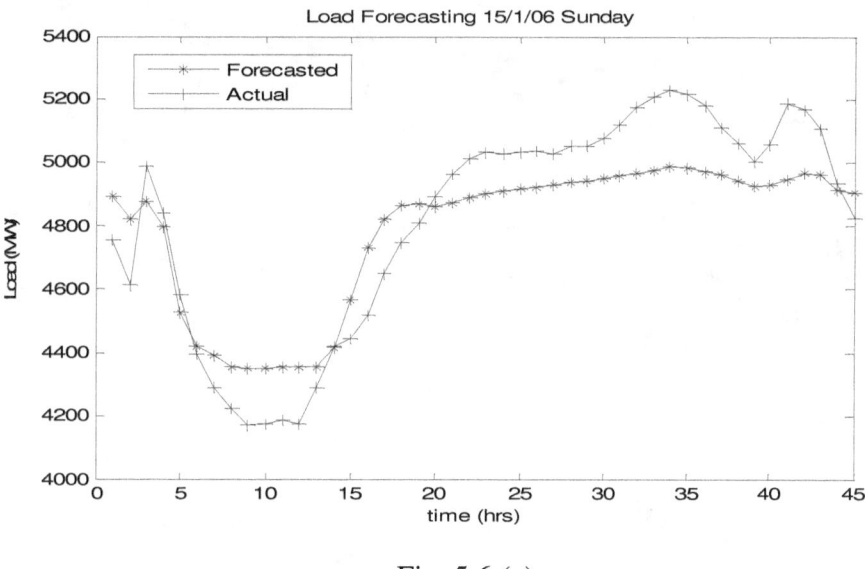

Fig. 5.6 (a)

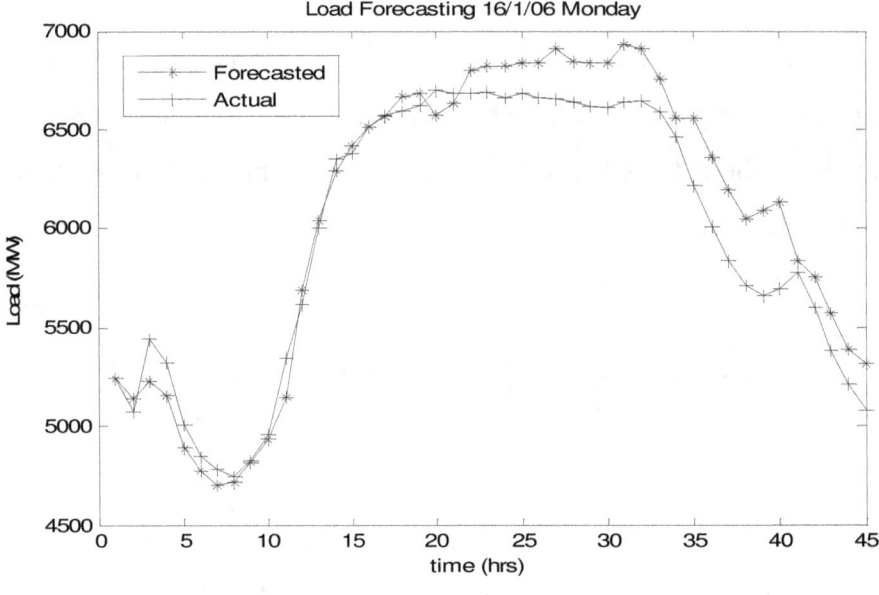

Fig. 5.6 (b)

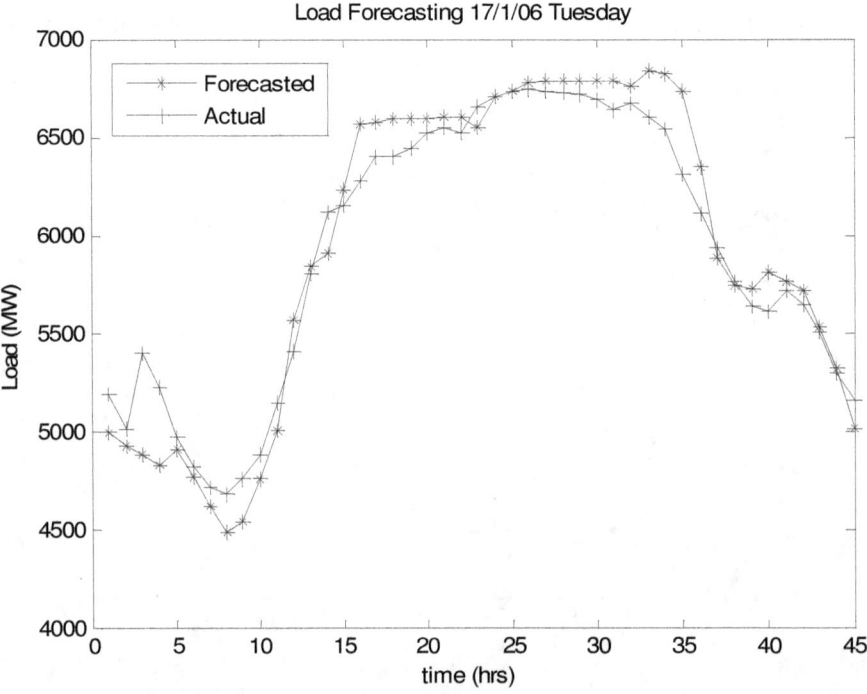

Fig. 5.6 (c)

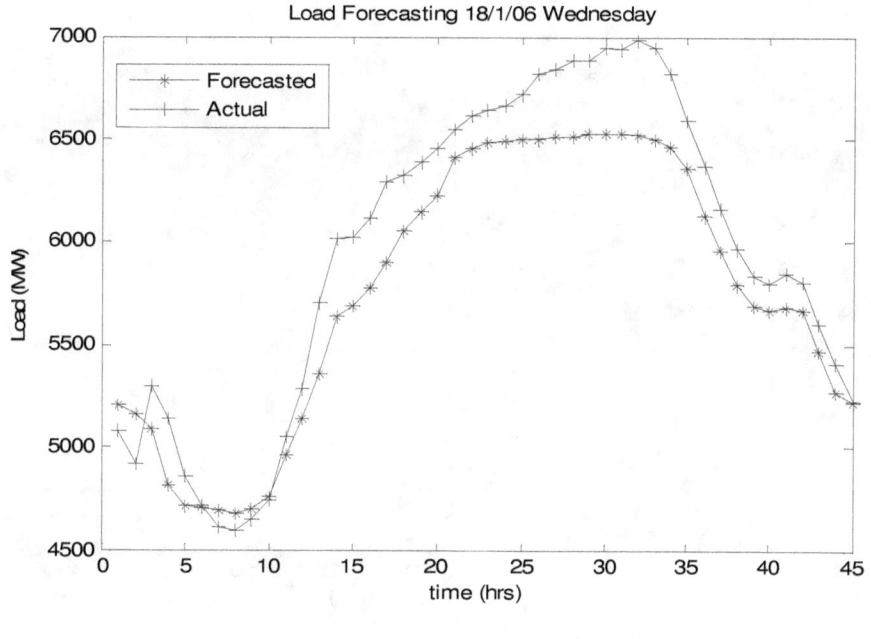

Fig. 5.6 (d)

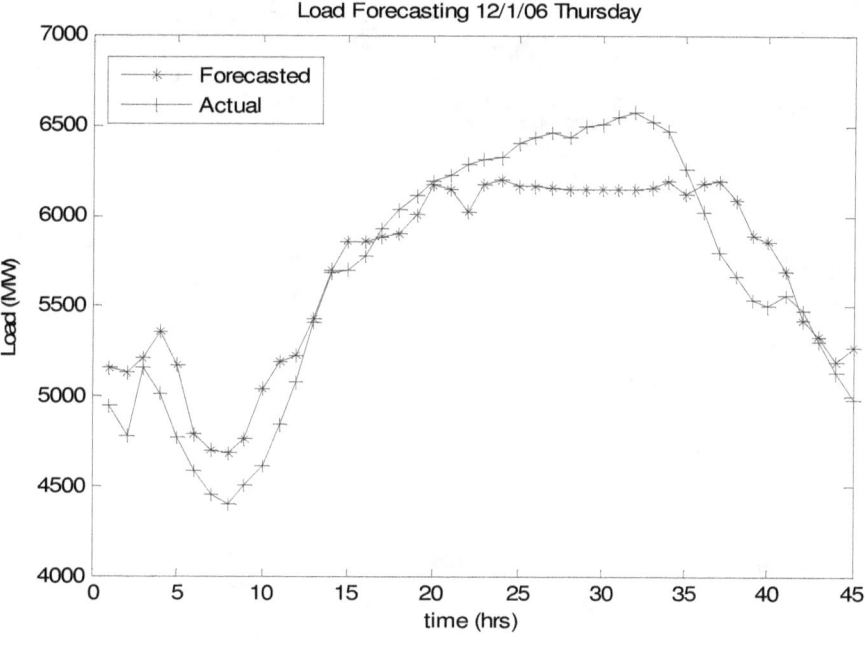

Fig. 5.6 (e)

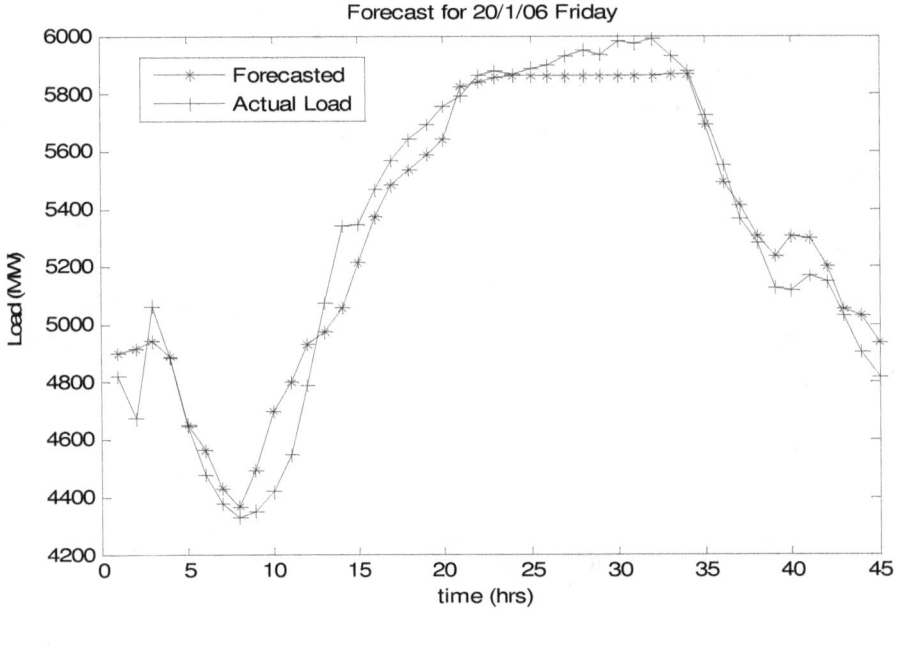

Fig. 5.6 (f)

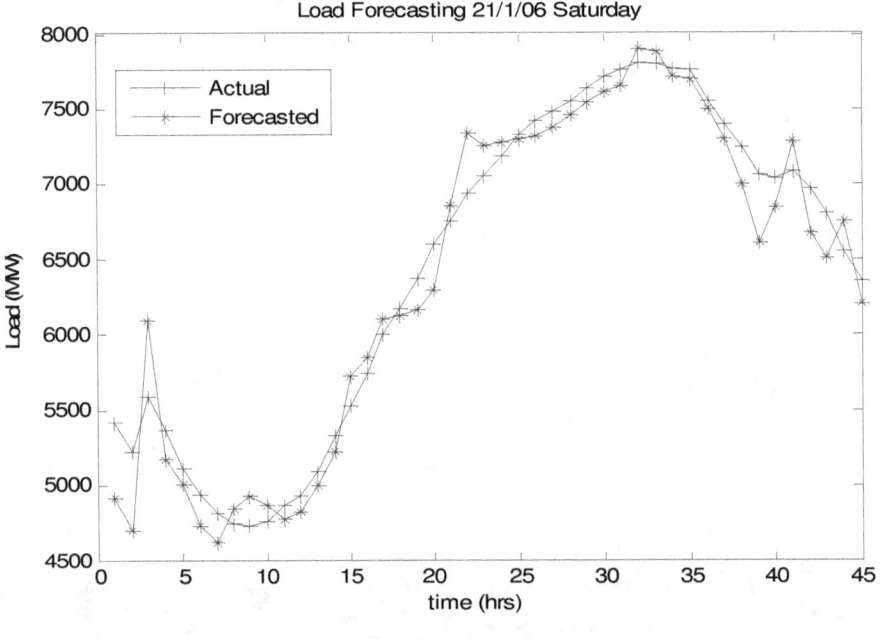

Fig. 5.6 (g)

Fig. 5.6: Load forecast using GRNN for Victoria market

Fig. 5.7 (a), (b), (c), (d), (e), (f) and (g) shows the maximum, minimum and MAPE errors in load forecasting of Victorian electricity market for each day of the week of 15-21 January 2006 with GRNN.

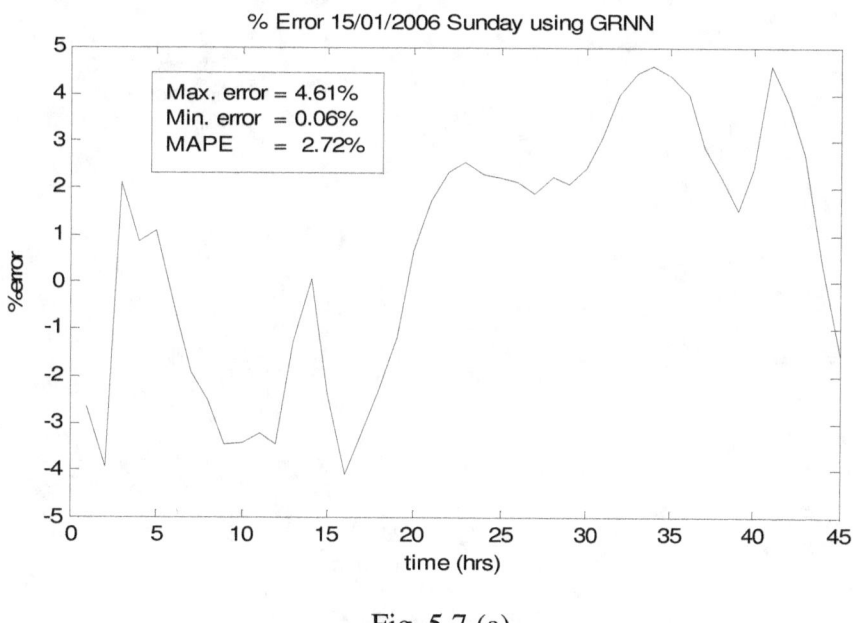

Fig. 5.7 (a)

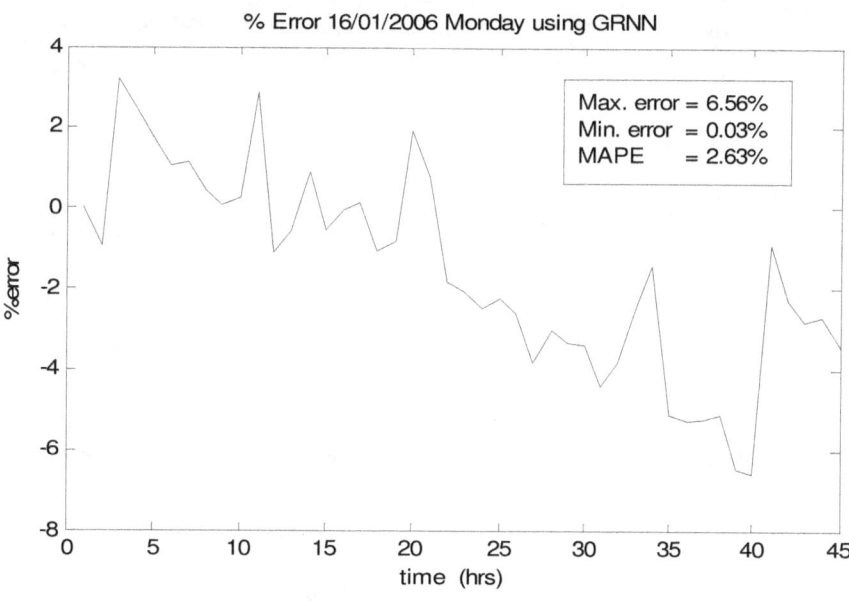

Fig. 5.7 (b)

These errors are different on different days of the week. The maximum error in load forecasting for Victoria market with GRNN varies from 1.2 % to 7.62 % and minimum error varies from 0.016 % to 0.16 % on different week days. The MAPE varies from 1.8 % to 4.0 %.

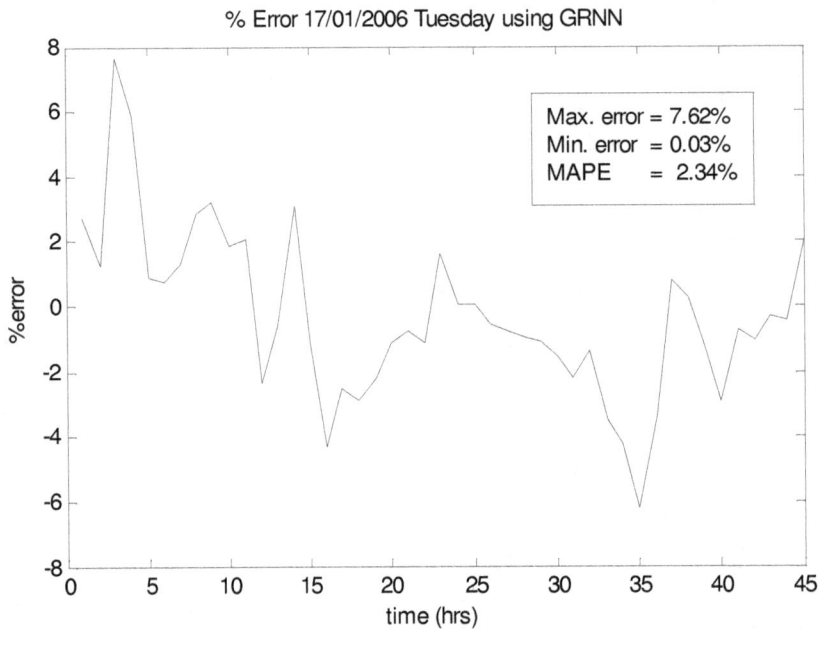

Fig. 5.7 (c)

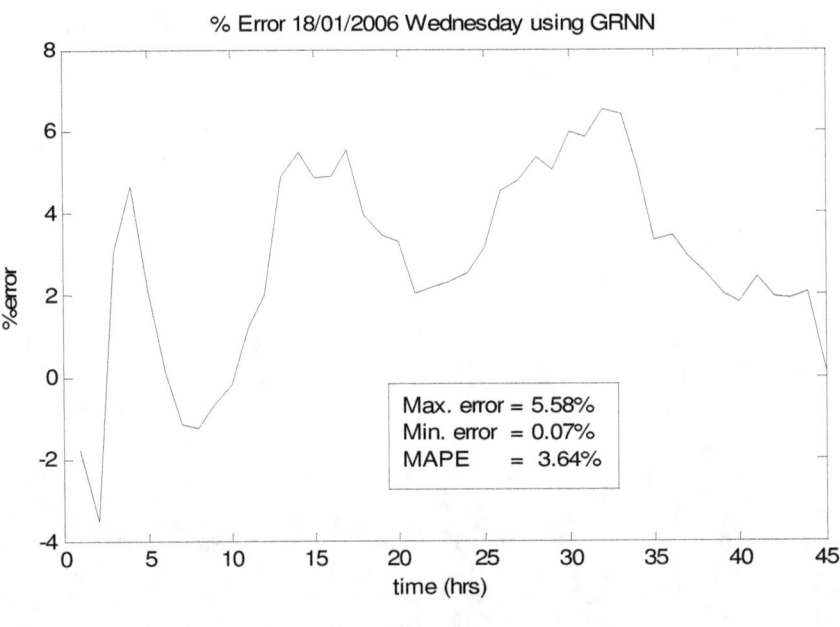

Fig. 5.7 (d)

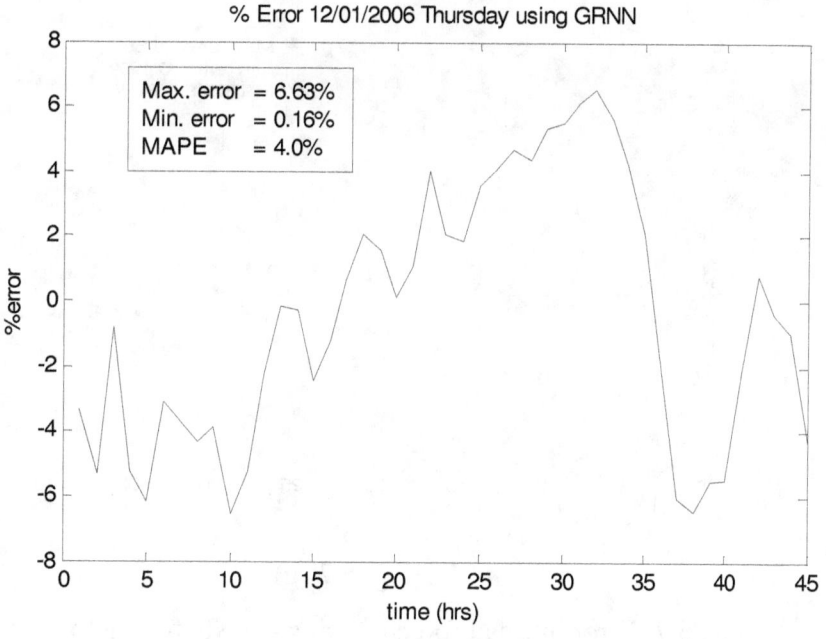

Fig. 5.7 (e)

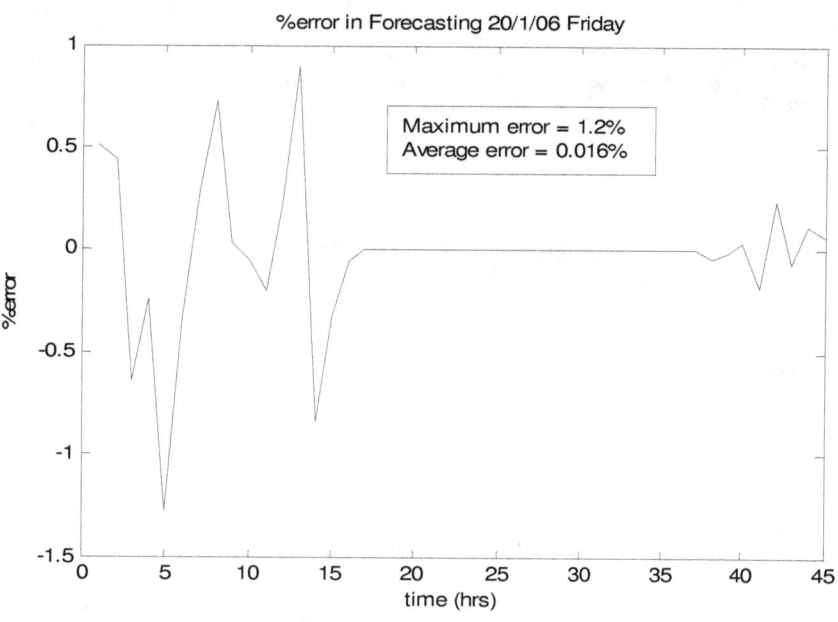

Fig. 5.7 (f)

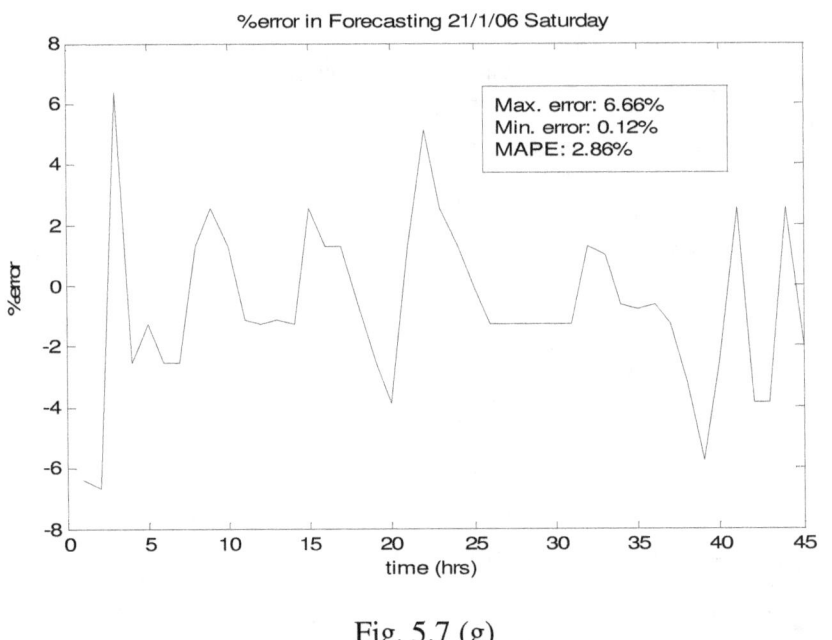

Fig. 5.7 (g)

Fig. 5.7: Error in Load forecast using GRNN for Victoria market

Fig. 5.8 (a) shows the actual and forecasted load of 72 hours (22-24 January 2004) in Victoria market using GRNN.

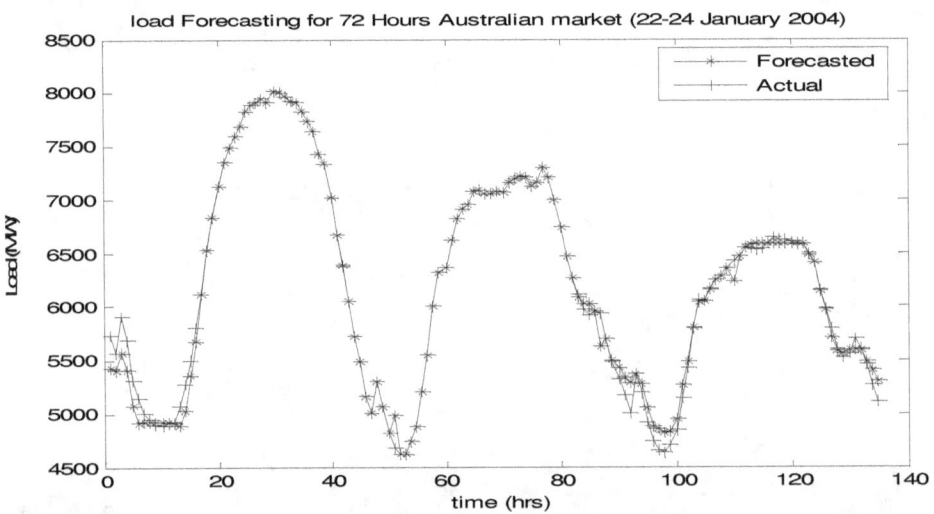

Fig. 5.8 (a): 72 Hours load forecast using GRNN for Victoria market

Fig. 5.8 (b) shows the error in load forecasting of 72 hours (22-24 January 2004) in Victoria market using GRNN.

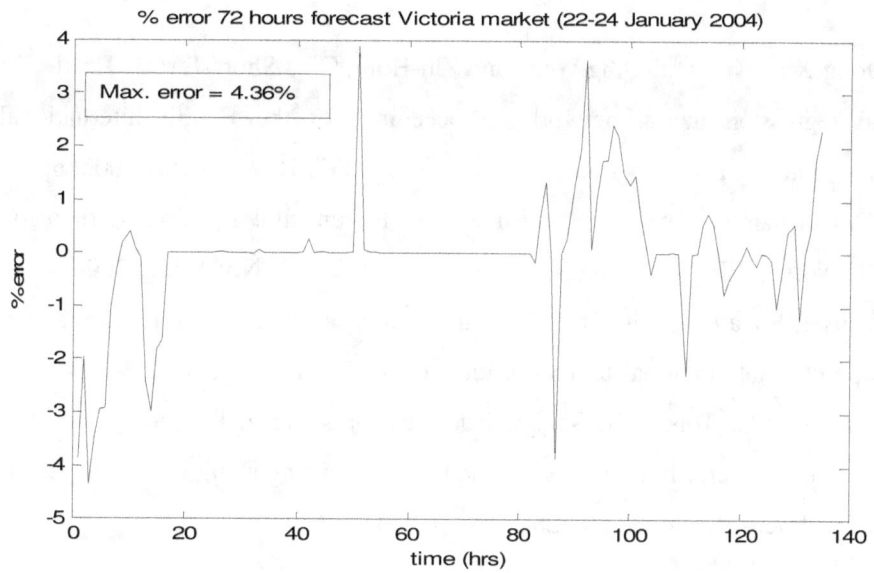

Fig. 5.8 (b): Error in 72 hours load forecast using GRNN for Victoria market

5.6 Summary

Load forecasting in the emerging electricity market plays a very important role for economic, secure and stable operation of power systems. ANN methods are found to be very attractive for load forecasting as these have the ability to handle the non-linear relationships between load and the factors affecting it directly. Volatile electricity price in power markets is a major input for load forecasting using ANN. In addition, only temperature (from weather variables) is used. An artificial neural network, known as generalized regression neural network (GRNN) has been used for explaining the short-term load forecasting of Victorian electricity market. The maximum errors in load forecasting for Victorian electricity market are 7.62 % and the maximum MAPE for Victorian electricity market is 4.0 % using GRNN. The simulation results show that ANN are effective as the forecasted load for Victorian electricity markets is very close to the actual load.

References

[1]   G. Gross and F.D. Galiana, "Short-term forecasting", Proc IEEE, Vol. 75, 1987, pp. 1558–1573.

[2]   Dong-Xiao Niu, Hui-Qing Wang and Zhi-Hong Gu, "Short-Term   Load   Forecasting using general regression neural network", Proceedings of the Fourth International Conference on Machine Learning and Cybernetics, Guangzhou, Vol. 7, 18-21 August 2005, pp. 4076-4082.

[3]   C. Constantinopoulos and A. Likas, "An Incremental Training Method for the Probabilistic RBF Network", IEEE Trans. on Neural Networks, Vol. 17, No. 4, July 2006.

[4]   Yousef, Rana el Hindi, Khalil, "Training radial basis function networks using reduced sets as center points", International Journal of Information Technology, Jan 2005

[5]   Paras Mandal, Tomonobu Senjyu, Naomitsu Urasaki and Toshihisa Funabashi,   "A   neural network based several-hour-ahead electric load forecasting using similar days approach", Electrical Power and Energy Systems, Vol. 28, 2006, pp. 367–373.

CHAPTER 6

Day-ahead Price Forecasting using Artificial Neural Network

6.1 Introduction

Price information is very important for market participants in the electricity markets. Future electricity price depends on several factors such as load, bids of generating companies and load entitles, gaming of market participants and several technical operating constraints [1]-[5]. Forecasting of electricity price accurately and efficiently has become more and more important for many activities, such as trading and risk management, to evaluate derivatives and devise hedging strategy. Once a good price forecast is available for the next day, large customers can derive a plan to optimize their own utility using the electricity purchased from the pool [6]. Like demand, electricity price has also its special characteristic as the seasonal variations at different level (daily, weekly and annual seasonality) [7]. Electricity may not be transported from one region to another because of existing bottlenecks or limited transportation capacity. Hence, price is local and differs among regions [4]. Furthermore, electricity price can rise to many times of its normal value showing great volatility.

PJM interconnection is one of the regional transmission organizations (RTO) that operate the largest competitive wholesale electricity market in the world [8]. The PJM power market is coordinated by an independent system operator (ISO) that ensures secure, economical and efficient operation as well as determining all locational marginal price (LMP) according to voluntary bids and bilateral transactions [9]. Forecasting LMPs help market participants, who bid into the spot price market, to determine the bidding strategy of their generators and provide better risk management [10]. In the PJM power market, it is observed that daily power demand curves have similar patterns, whereas the daily LMP curves are volatile. The LMP and load values are low at night. In general, loads and prices in the wholesale markets are mutually intertwined activities [11].

Several methods have been proposed in the literature for forecasting electricity prices with different time horizons such as time-series [12], artificial neural networks (ANN) [4], [9], [13]-[15], wavelet transform and ARIMA models [16], agents based simulations [11], input- output hidden Markov models (IOHMM) [17], etc. Among the various methods, ANNs have received much attention in recent years for modelling complex non-linear relationships [9], [11], [18]. ANNs are used for solving the short-term load forecast problems in reference [12], [19]-[21]. ANNs are also used to forecast system marginal price (SMP) [14], market clearing price (MCP) [22], [23] and market clearing quantity (MCQ). In this chapter, generalized regression neural network (GRNN) having a radial basis layer and a special linear layer, has been used to forecast day-ahead electricity price using PJM electricity market data [24]. The effectiveness of the proposed method has been tested and results show that the GRNN is able to forecast accurate future prices.

6.2 Day-ahead Price Forecasting

6.2.1 Characteristic of Electricity Prices

Electricity is a special commodity, which differs in several ways from other commodities. The impossibility of storing bulk amount of electrical energy to deliver in later periods and the necessity of assuring a constant equilibrium between supply and demand from stability point of view, associated with transmission congestion that may prevent free exchange among control areas, make the electricity markets to be operated differently. These difficulties introduce special characteristics to the electricity prices and in particular, to the spot prices.

In most of the competitive electricity market, price series is characterized by high frequency, high volatility and non-stationary having multiple seasonality effect, calendar effect on weekend and holidays and outliers' presence. Irrational bidding behaviour by the market participants (producers and consumers) leads to high volatility of electricity prices. Price volatility is a measure of instability in future prices or uncertainty in future price movements [25].

The choice between uniform and pay-as-bid pricing for electricity auctions has been one of most important issues in newly deregulated electricity markets. Under the uniform pricing structure, the marginal bid block sets the market clearing price (MCP). However, in a pay-as-bid (discriminatory) pricing structure, every qualified generator is paid what they offer [26]. Uniform pricing scheme is the most commonly adopted structure of the electricity markets around the world. Generally, when there is no transmission congestion, and system losses are ignored, MCP is the same for the entire system, which is also known as system marginal price (SMP). However, when there is congestion, the zonal market clearing price (ZMCP) or the locational marginal price (LMP) could be employed. ZMCP may be different for various zones, but it is the same within a zone. LMP can be different for different buses [27]. The pricing mechanism can affect the competition, efficiency, consumer surplus and total revenue of the players in the electricity markets.

6.2.2 Factors Affecting Electricity Prices

The electricity market price, which is determined by the ISO with the available bids from supply and demand sides, is influenced by many factors. Demand, including both historical and forecasted demand as well as accuracy of forecasted demand, and supply are direct factors influencing electricity price. Errors in demand forecasts and demand scheduling are one of the major reasons of price swings. Exercising the market power by dominant generators could cause volatility, in addition to raising prices over competitive levels. When demand-forecasting error exists along with market power, the volatility in price could be more severe. The fuel (mainly natural gas) used by generating units to produce electricity is a volatile commodity.

Availability and irregularity of relatively inexpensive generation facilities (e.g., hydro and nuclear) also affects the electricity price. When network congestion occurs, the lowest-cost electricity cannot be delivered to all the loads due to transmission bottlenecks. In congested network, the higher-cost generation units may have to be dispatched to meet the load and hence, the price of energy in a constrained area may be higher than an unconstrained area. Other factors such as production cost, planned outages for maintenance schedule and forced outages for unplanned maintenance of generators or transmission lines, reserve margin, etc influence the market price in direct and indirect ways. Moreover, the structure and the management rules of any specific electricity market may also affect the electricity price.

The factors affecting the electricity prices are categorized as given below:

- **Structural factors** including market design and types of production units,
- **Behavioural factors** including bidding/pricing strategy and time index,
- **Operational factors** including availability of hydro generation, network congestion and margin, reserve margin, forced outage rate and maintenance schedule,
- **Historical factors** including electricity prices, demand and
- **External factors** such as weather, fuel prices and supply.

Some of the key issues and challenges in electricity price forecasting are:

- Consideration of distributed generation (DG) in the price prediction.
- Most electricity systems lack storage for all practical purposes.
- The electric system is subject to the confluence of unusually high demand, unexpected generator outages and transmission de-ratings.
- Instantaneous demand and supply imbalances subjected the system to unusual stress.
- Emission allowances (EA).
- Past conditions are unlikely to be repeated in any consistent manner useful in forecasting.

- Power system networks frequently get congested due to increased transaction and limited transmission system expansion, which raise the market price.

- Forecasting strategic behaviour of suppliers is a key concern in price forecasting.

- Several markets run parallel to the energy market such as ancillary market, transmission market.

- Analysis of economic impact of price forecast errors for short-term scheduling is to be analysed.

6.3 Price Forecasting Methods

Market simulation methods, which, generally makes a large amount of hypotheses, consider all generators' operation cost with the generators' limits and transmission constraints. This approach is suitable for forecasting long-term price forecasting. Statistical methodology, which is, based on the assumption of historical price characteristics, can be categorized to the time series models, intelligent system methods and volatility analysis. Many of these methods are the load forecasting and especially Short Term Load Forecasting (STLF) methods.

6.3.1 Time Series Models

Time-series models provide a trade-off between underlying price behaviour and accurate forecasting. Contreras et al. [28] have developed Auto Regressive Integrated Moving Average (ARIMA) model to forecast electricity market prices of the Spanish and Californian markets. Multivariate Dynamic Regression (DR) and Transfer Function (TF) models have been applied for forecasting the Spanish and California market prices [29] and the PJM market prices [30].

DR relates current price to the values of past prices and past demands whereas TF relates current price to past prices, past demands and past errors. Early applications of the time series models in the power system were related to STLF [31]. A Bayesian-based classification method combined with an AR model is presented in [32] to predict the discrete Probability Density Functions (PDFs) of the MCPs.

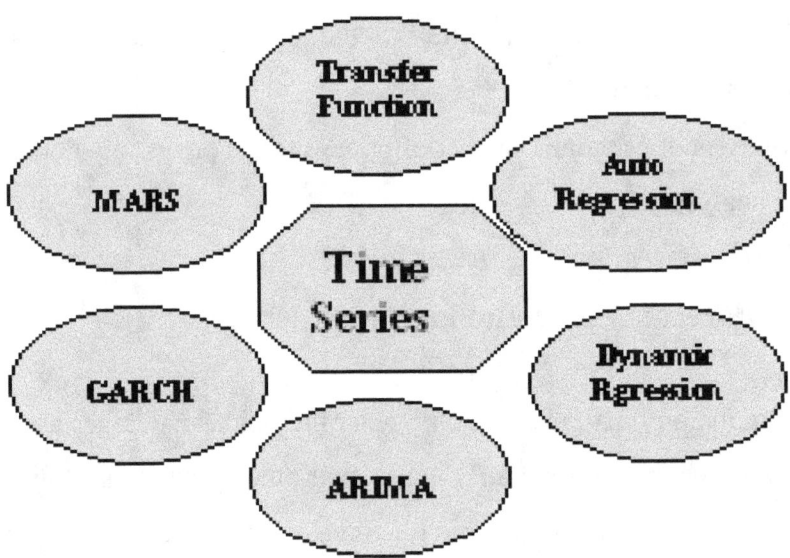

Fig 6.1: Time series methods

The non-linear MARS (Multivariate Adaptive Regression Splines) technique has been applied to forecast hourly energy price [33]. Generalized Autoregressive Conditional Heteroscedasticity (GARCH), uses past variances and past variance forecasts to forecast future variance and has been applied to forecast day-ahead electricity price for Spain and California market [34].

6.3.2 Artificial Intelligence System Models

Artificial Intelligence (AI) techniques include Artificial Neural Networks (ANN), fuzzy logic, and their combinations as shown in fig. 6.2. A typical ANN for electricity price forecasting is feed forward Multi-Layer Perceptron (MLP) model with Back Propagation (BP) training algorithms (gradient descent) [35]. A single neural network with traditional learning algorithms may not be suitable for complex nonlinear mapping function of the price signal and cascaded architecture of multiple ANNs [36] and committee machine [37] replace the single neural network. Radial Basis Function Neural Network (RBFNN) [38] and Recurrent Neural Network (RNN) [39] have also been proposed for forecasting due to several advantages.

To take care of the high-frequency changes of the MCP, fuzzy model [40] has been applied to forecast the possible ranges of variation in the electricity price. L. Hongjie et al. [41] have proposed Dynamic Fuzzy System (DFS) to forecast MCP of California power market. Also a kind of Extended Kalman Filter (EKF) [42] and an Input/Output Hidden Markov Model (IOHMM) [43] respectively has been applied to New England and Spain electricity market respectively.

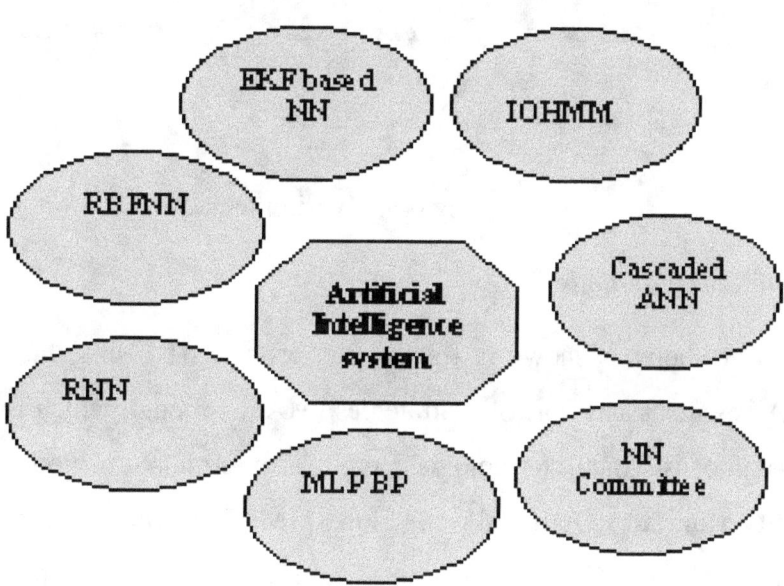

Fig 6.2: Artificial Intelligence methods

6.3.3 Hybrid Models

Hybrid models are shown in fig. 6.4 and discussed in this section. A new Fuzzy NN (FNN) with higher learning capability than ANNs has been proposed to forecast electricity prices [44] to solve the problem like inadequacy of a single Neural Network (NN) to construct a global model for the MCP signal. Many researchers have applied wavelet transformation as a pre-processor to decompose the ill-behaved price series into better behave consecutive series and then forecasting model like ARIMA [45] and ANN [46] have been applied for price forecasting.

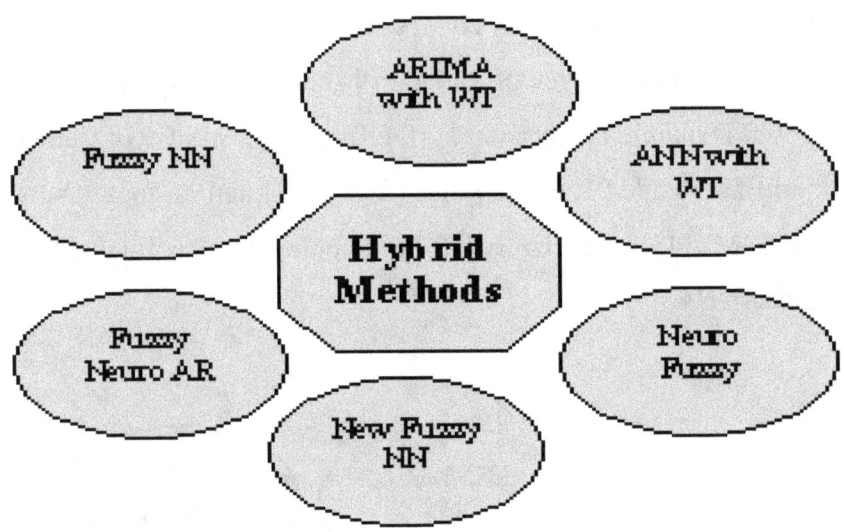

Fig 6.3: Hybrid methods

6.3.4 *Simulation Methods*

Simulation methods as shown in fig. 6.4 can provide detail insights into the price curve. Market Assessment and Portfolio Strategies (MAPS) model, which is based on LMP, is a transmission-constrained chronological production simulation model. A structural Multi-commodity Multi-area Optimal Power Flow (MMOPF) is also a product simulation model that performs Monte Carlo simulation to take into account all major drivers, including participants' bidding behaviour, can provide reliable prices and realistic option values.

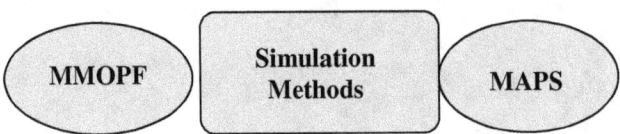

Fig 6.4: Simulation methods

6.3.5 Volatility Analysis

Different approaches for measuring wholesale electricity market price volatility have been proposed. Alvarado et al. [47] have modelled electricity prices using a frequency domain analysis to separate the cyclic component of price. A study by Benini et al. [48] shows in Spain, California, and PJM electricity markets that the price volatility is strongly connected to the installed generation capacity. T. Mount [49] argued that the use of a uniform price auction for electricity markets worsens price volatility and a pay-as-bid price auction is a better alternative.

6.4 Forecasting Framework

The time framework to forecast the day-ahead market prices in most of the electricity markets is explained and illustrated in fig. 6.5. The market price forecasts for day, d, are required on one day ahead (d-1), typically at hour h_b (around 10 am). On the other hand, data concerning results for day (d-1) are available on day (d-2) at hour h_c (around noon). Therefore, the actual forecasting of market prices for day d can take place between hour h_c of day (d-2) and hour h_b of day (d-1). Therefore, to forecast prices for day d, price data up to hour 24 of day (d-1) are considered to be known [4].

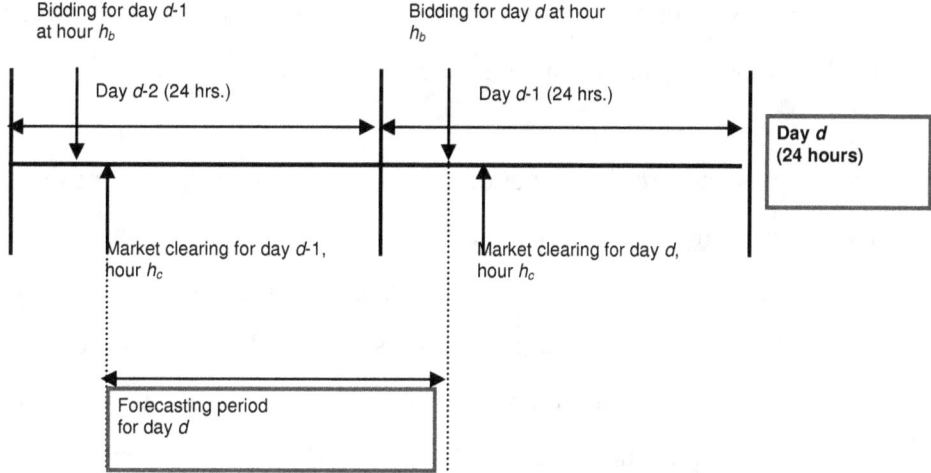

Fig. 6.5: Time framework to forecast market prices for day, d

6.5   Type and Structure of GRNN

The generalized regression neural network (GRNN) having a radial basis function layer and one special layer is used, here, as a forecasting tool in this application. GRNN is made-up of 4 layers and these are input layer, pattern layer, summation layer and output layer, respectively. The pattern layer is also called latent regression layer. Each unit is correspondent to a training sample. Gauss function $e^{-d(x_0, x_i)}$ is regarded as activate nuclear function, x_i is the canter vector of nuclear function of each unit and there are total n units. Summation layer include two units, the first unit calculates weighted sum of each unit output of pattern layer and the weight is value of y_i of every training sample that can be regarded as molecule. The second layer calculates the sum of the outputs of each unit of pattern layer and can be regarded as denominator. It is called the unit of denominator. The units of output layers divide the unit of molecule and denominator of summation layer to get estimation value y.

6.6 Input Selection for GRNN

Price-load relationship is relatively stable over shorter periods of time. It is clear from the scatter plot shown in fig. 6.6 that there is a strong correlation between load and price. Hence, historical load has been considered as input to neural network.

Fig. 6.7 shows the cross correlation plot of daily load and electricity price time-series against hourly time lag in PJM electricity market. It shows that time series at one hour before the forecasted hour (d-1) has the highest peak and hence, to be considered as input for price forecasting. Historical price has also been taken as input to the neural network. It is clear from the autocorrelation plot of electricity price time-series in PJM electricity market of fig. 6.8 that time series at (d-1) is highly correlated and hence, price at one hour before the forecasted hour (d-1) is to be considered as input for price forecasting.

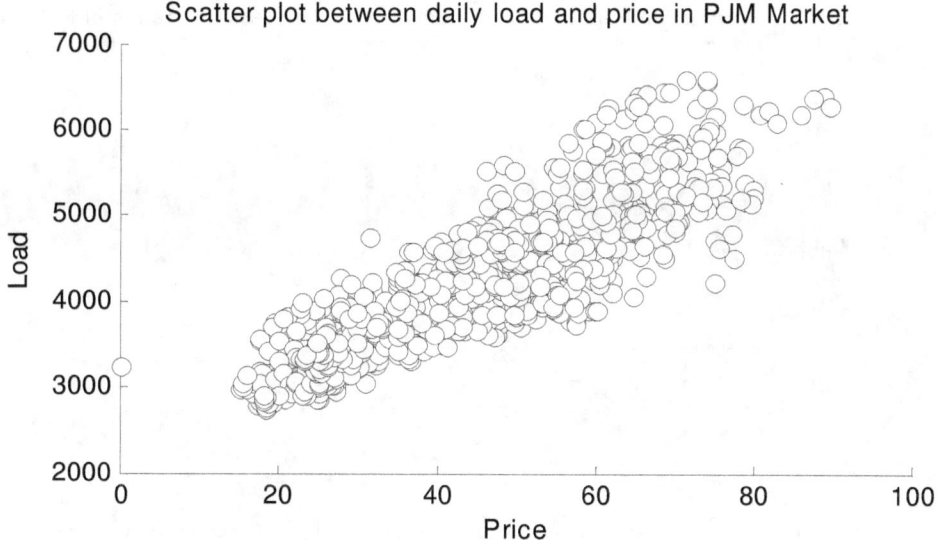

Fig 6.6: Scatter plot between daily load and price in PJM market

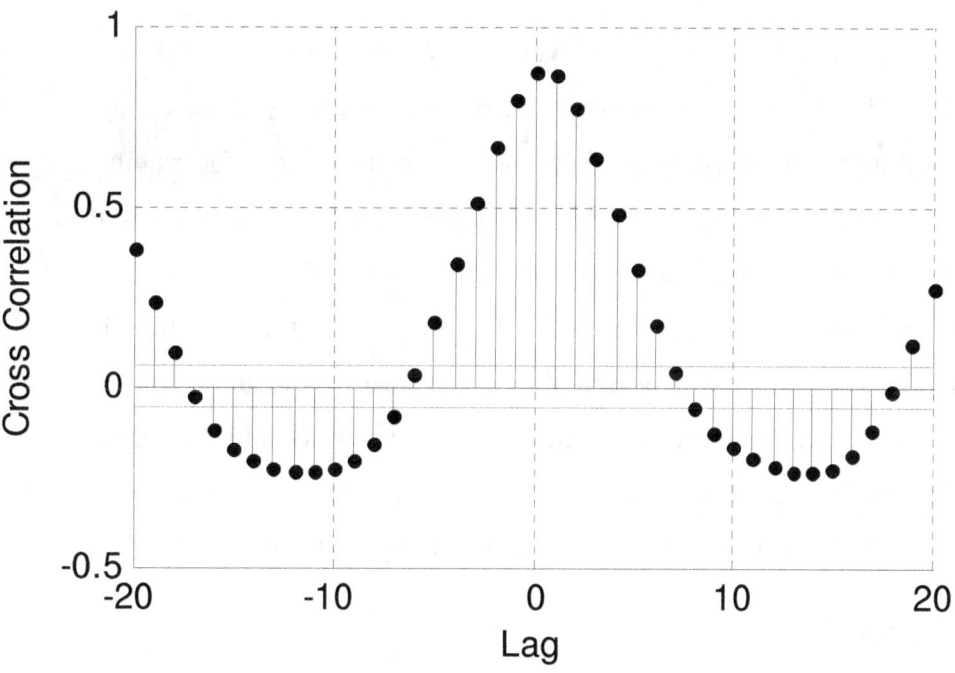

Fig 6.7: Cross correlation between daily loads and price time-series

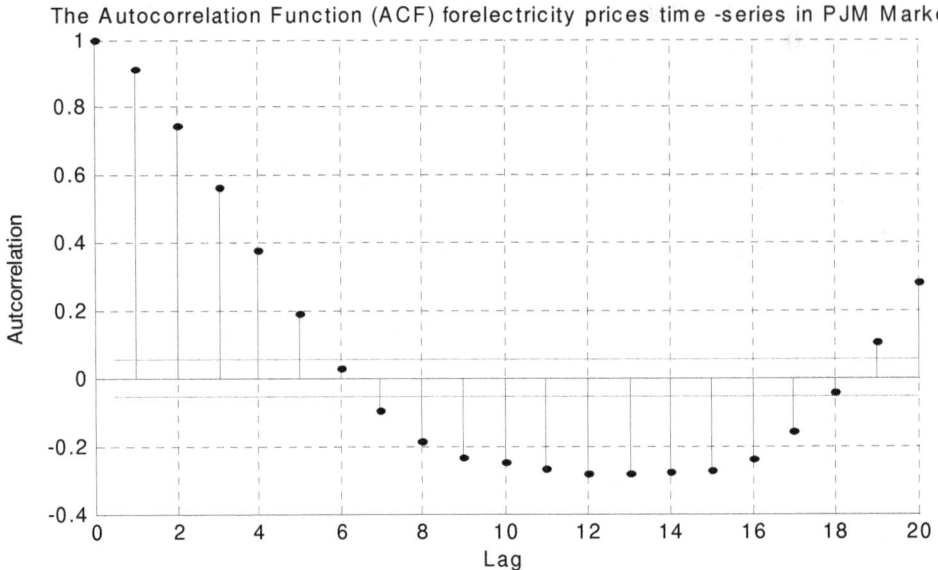

Fig 6.8: Auto correlation for Electricity price time-series

6.7 GRNN Training Data and Algorithm

Publicly available data from the web sites of PJM electricity market have been used to forecast the day-ahead electricity price of PJM electricity market [50], [51]. The data are divided into several windows where most of them are used for training and the remaining data are used for testing the GRNN. More precisely, for each month, the first week, second week and the third week are used for training, while the fourth week is left for testing the GRNN. Training was done for all the data windows at the same time i.e. the same GRNN is trained to be used at any time during the year. All inputs and outputs are normalized before training. Cross correlation between load and price and auto-correlation of price time series are found and only those load and price inputs are considered which are the best correlated. The inputs to the ANN as shown in fig. 6.9 are:

- $H(k)$ hour indicator
- $D(k)$ day indicator
- $L(k-1)$ Load of previous day at hour k
- $P(k-1)$ Price of previous day at hour k

- $T(\max)$ Maximum temperature of the day

- $T(\min)$ Minimum temperature of the day

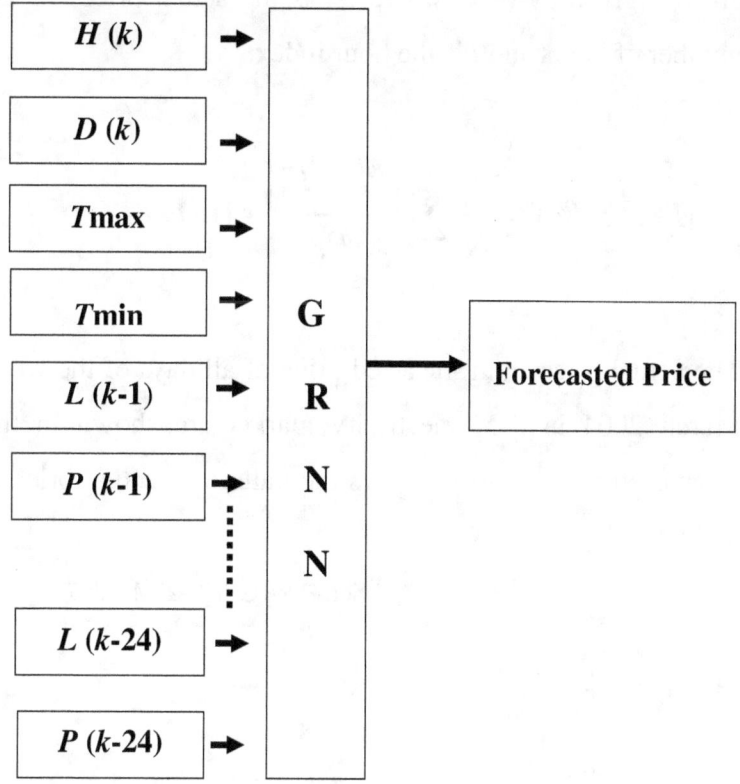

Fig. 6.9: The ANN model used for price forecasting

6.8 Simulation and Results

Simulation is carried out using MATLAB software. First, the developed GRNN model is trained using set of input/ output data. The performance of the trained GRNN model for forecasting day-ahead electricity price is tested using windows of data that are not included in the training set. Due to its special features as explained in the previous sections of this chapter and also in Chapter 2, the algorithm resulted in a very fast training and the error is significantly reduced to very low value.

For more accurate evaluation of the ANN performance, an average of the absolute error over a period of time, known as mean absolute percentage error (MAPE), is used for an overall evaluation and comparison with other techniques. The mean absolute percentage error is given by (6.1) where P_A is the actual price, P_F is the forecasted price, N is the number of hours and i is the hour index.

$$MAPE(\%) = \frac{1}{N}\sum_{i=1}^{N}\frac{\left|P_F^i - P_A^i\right|}{P_A^i} * 100 \qquad (6.1)$$

The actual price and forecasted price of all days of the week ranging from 22 March to 28 March 2004 in PJM electricity market are shown in the fig. 6.10 to Fig. 6.16. The forecasted price almost follows the pattern of actual price on all week days except Friday and Saturday.

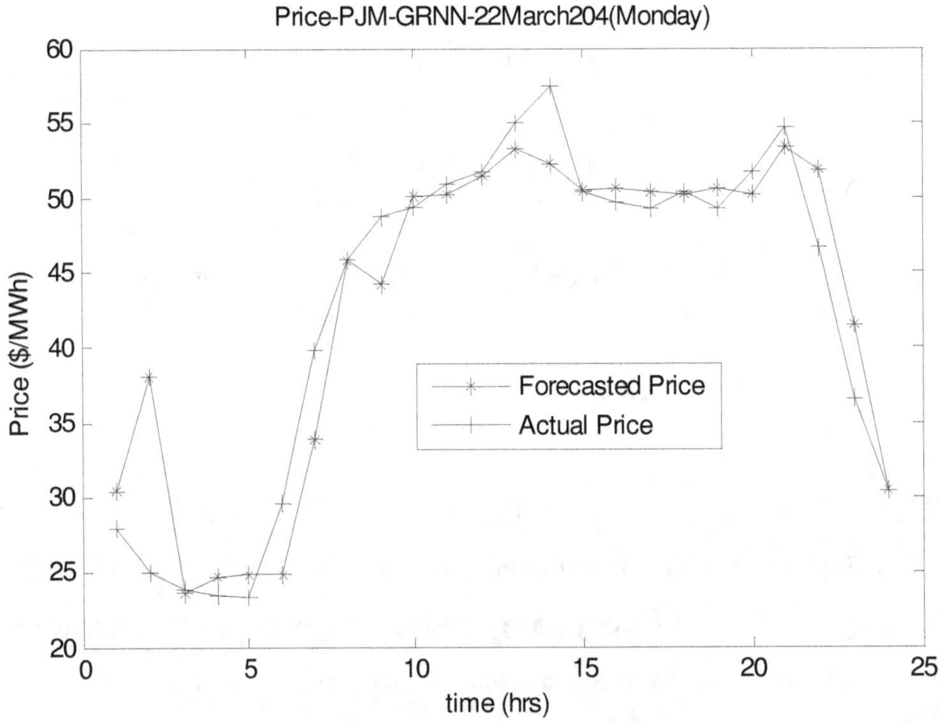

Fig. 6.10: Price forecast of Monday

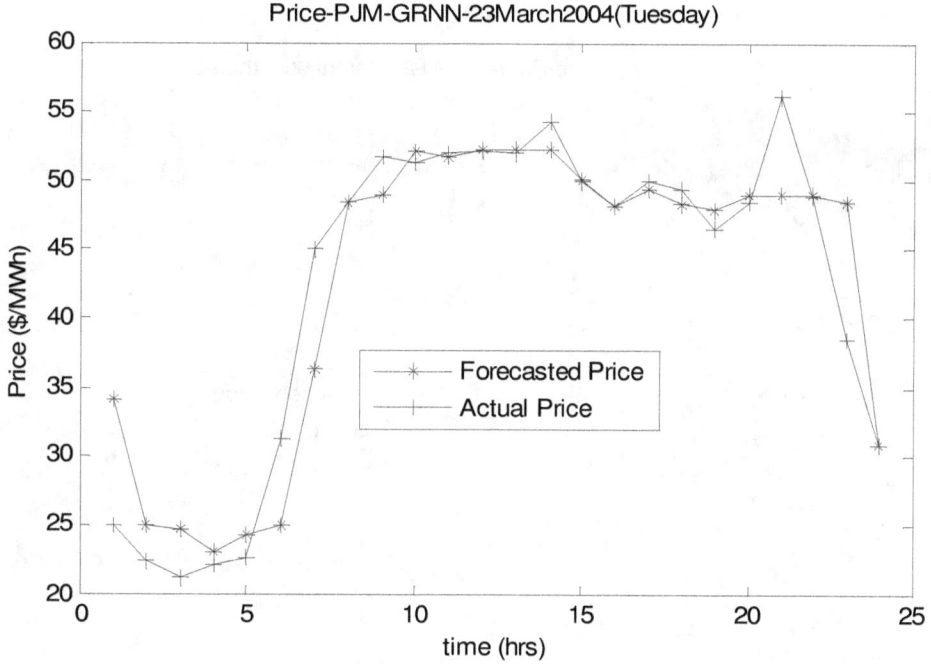

Fig. 6.11: Price forecast of Tuesday

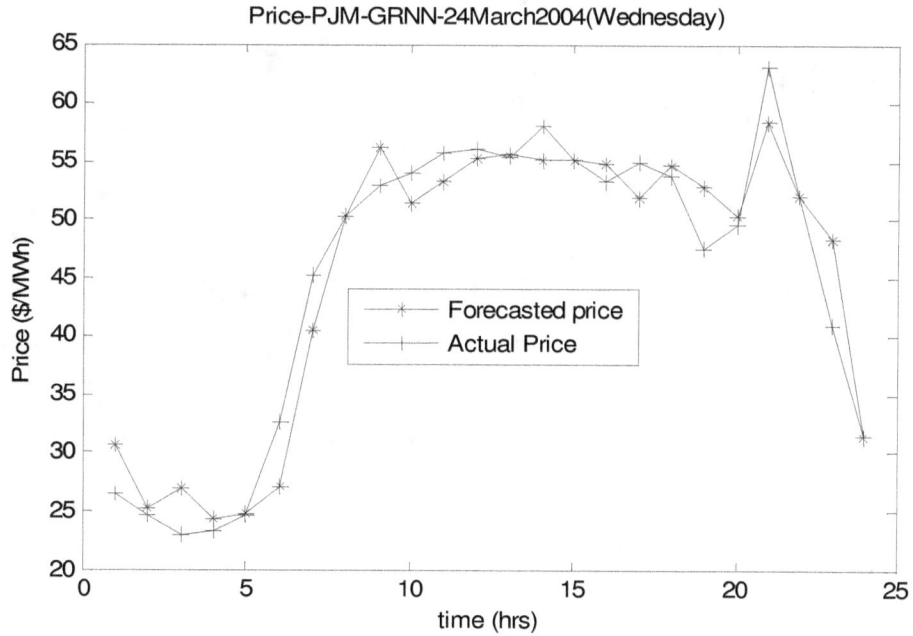

Fig. 6.12: Price forecast of Wednesday

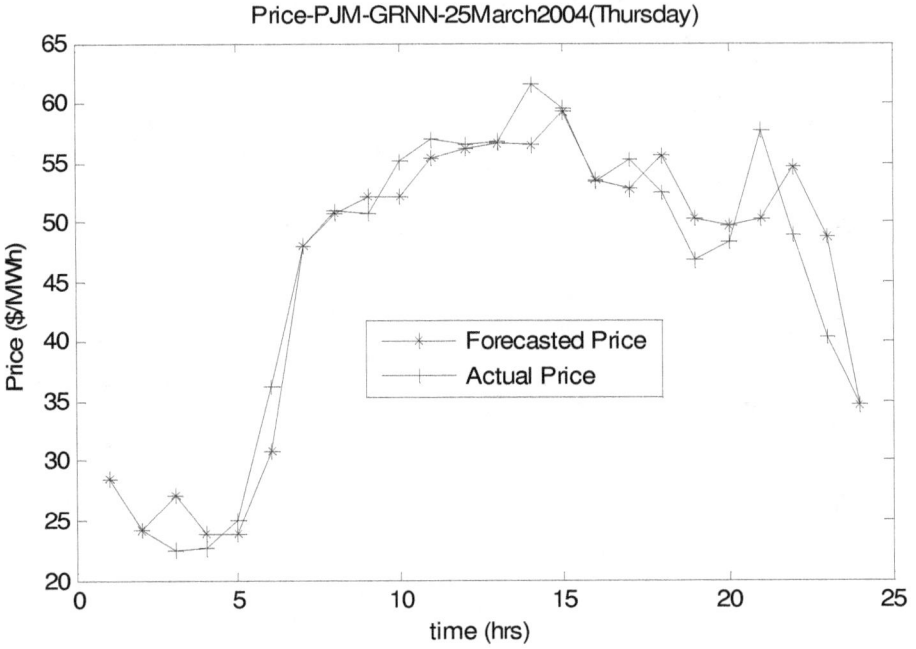

Fig. 6.13: Price forecast of Thursday

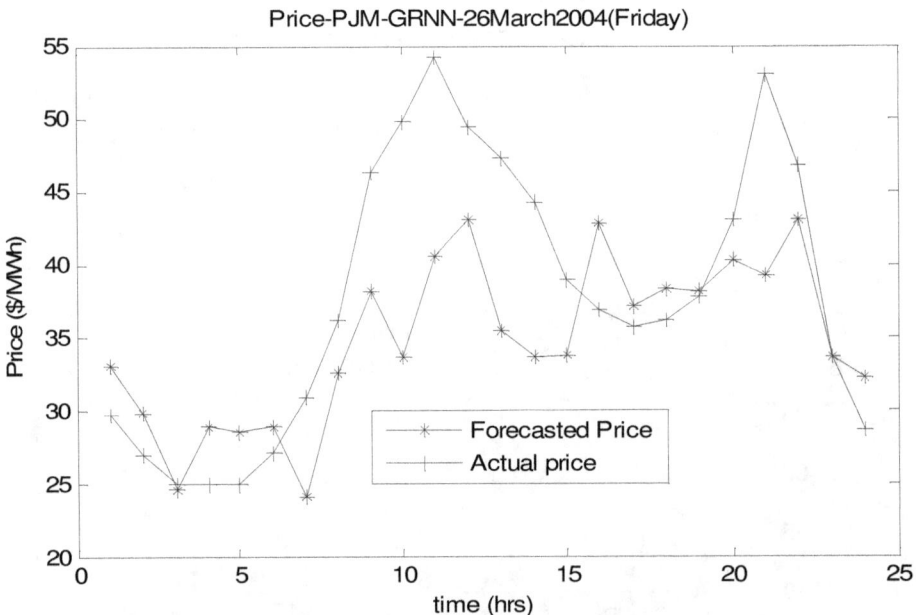

Fig. 6.14: Price forecast of Friday

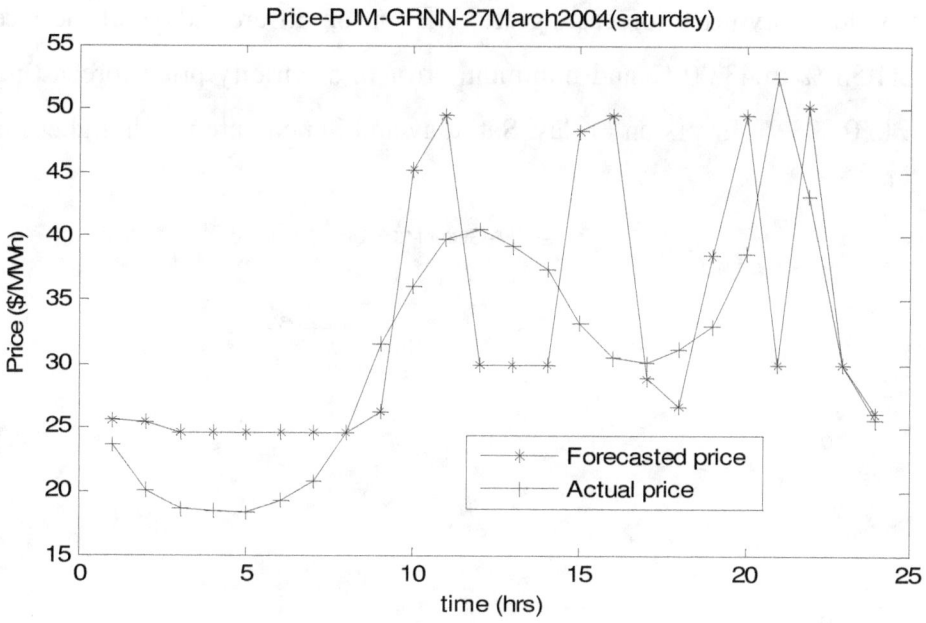

Fig. 6.15: Price forecast of Saturday

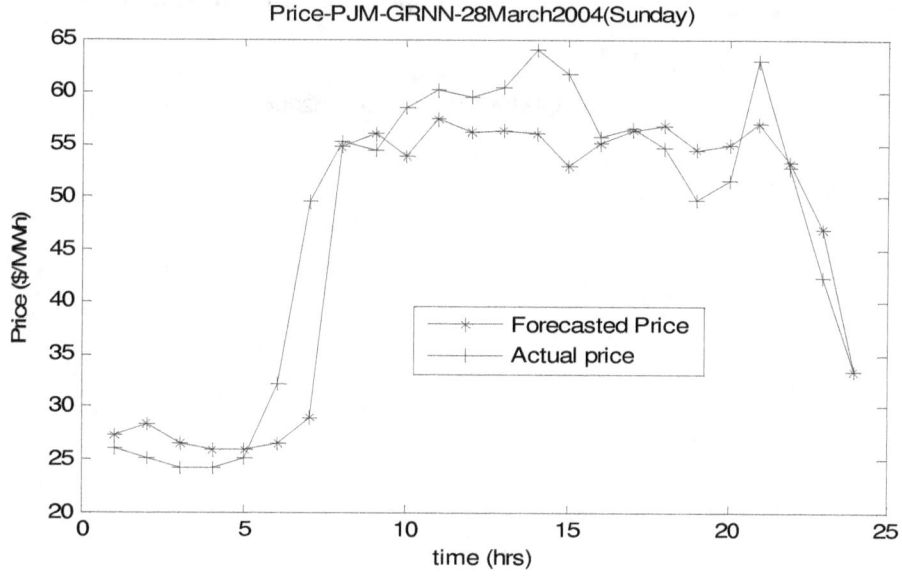

Fig. 6.16: Price forecast of Sunday

Fig. 6.17 to Fig. 6.23 shows the errors in electricity price forecasting on all days of the week ranging from 22 March to 28 March 2004 in PJM electricity market. The maximum error in electricity price forecasting are different on different days of the week and varies from 11.53 % to 43.00 % and minimum error in electricity price forecasting varies from 0.0 % to 0.003 %. Errors on Friday, Saturday and Sunday are much higher than other days of the week.

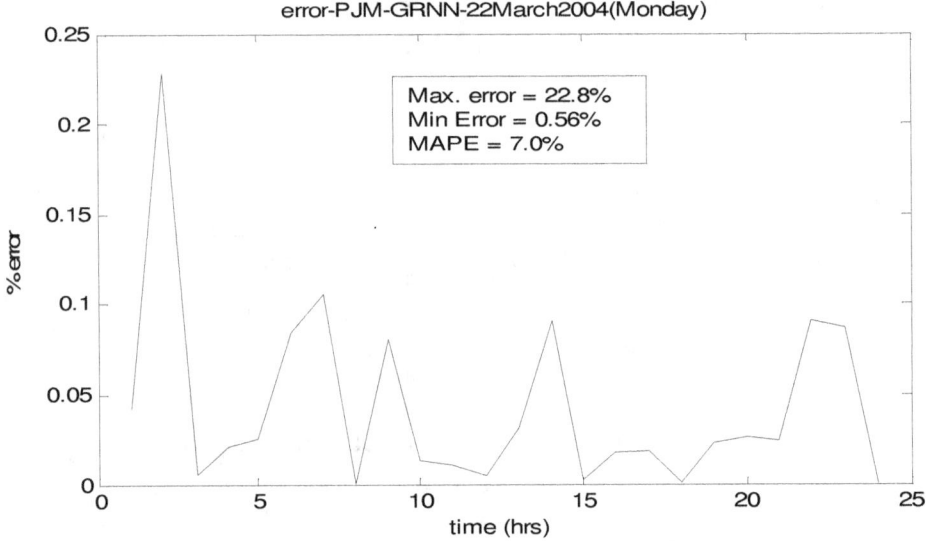

Fig. 6.17: Error in price forecast of Monday

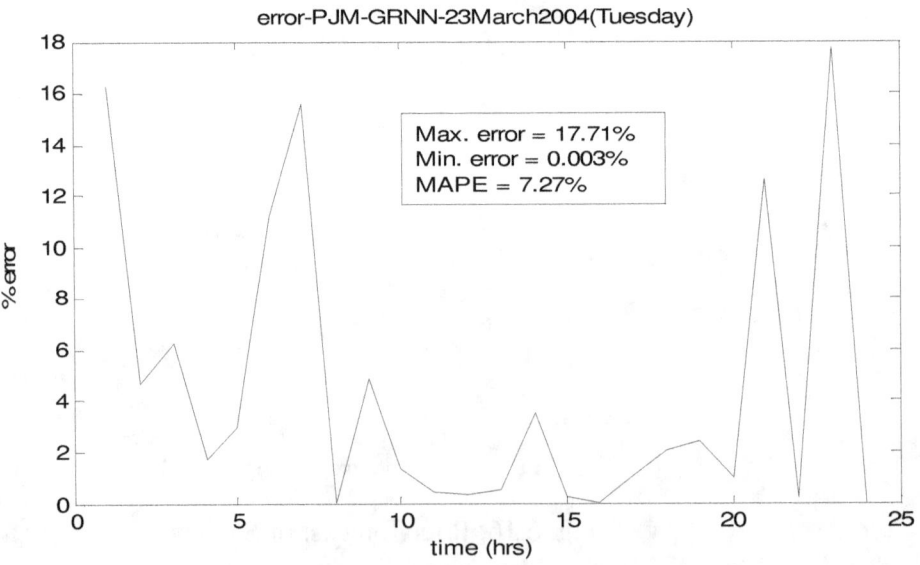

Fig. 6.18: Error in price forecast of Tuesday

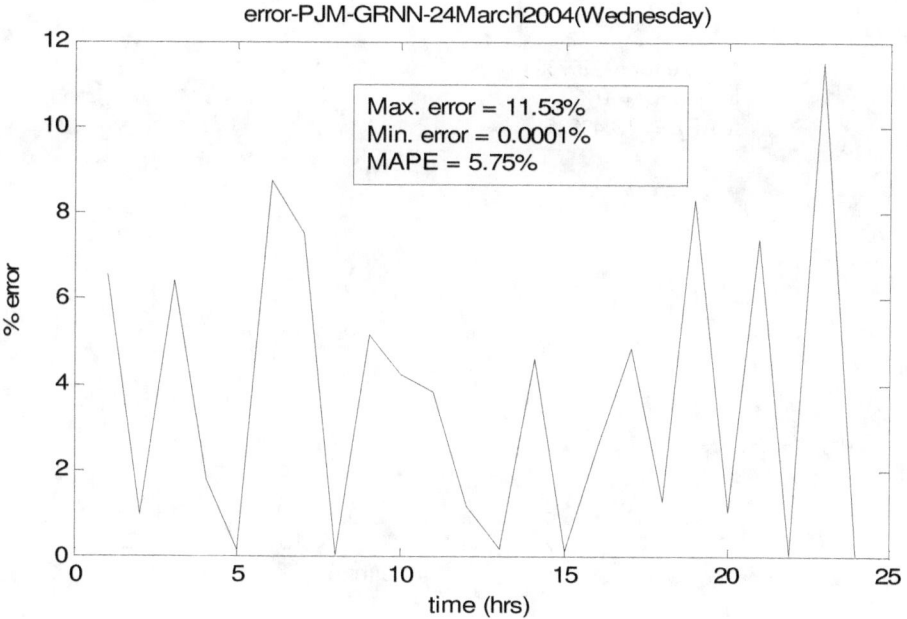

Fig. 6.19: Error in price forecast of Wednesday

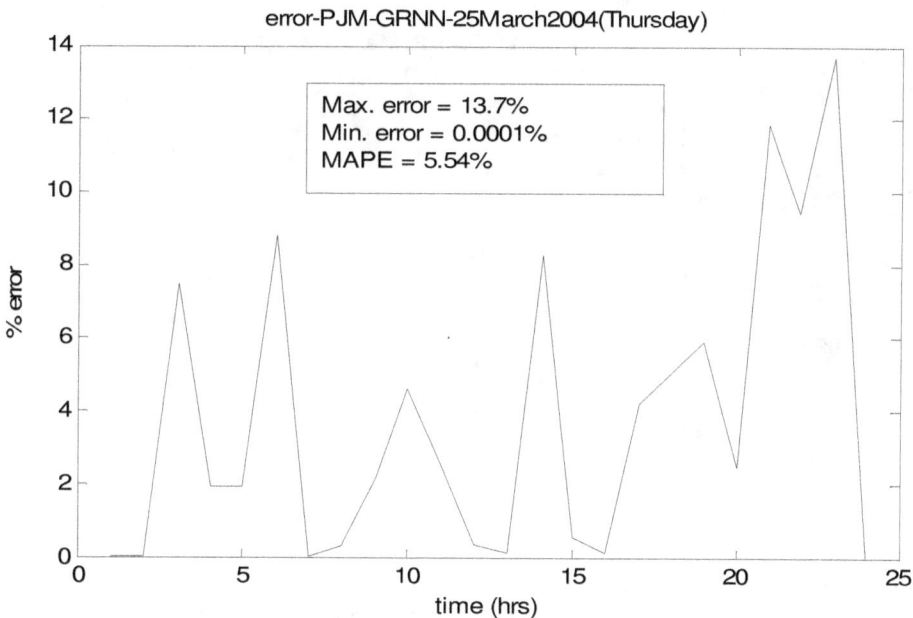

Fig. 6.20: Error in price forecast of Thursday

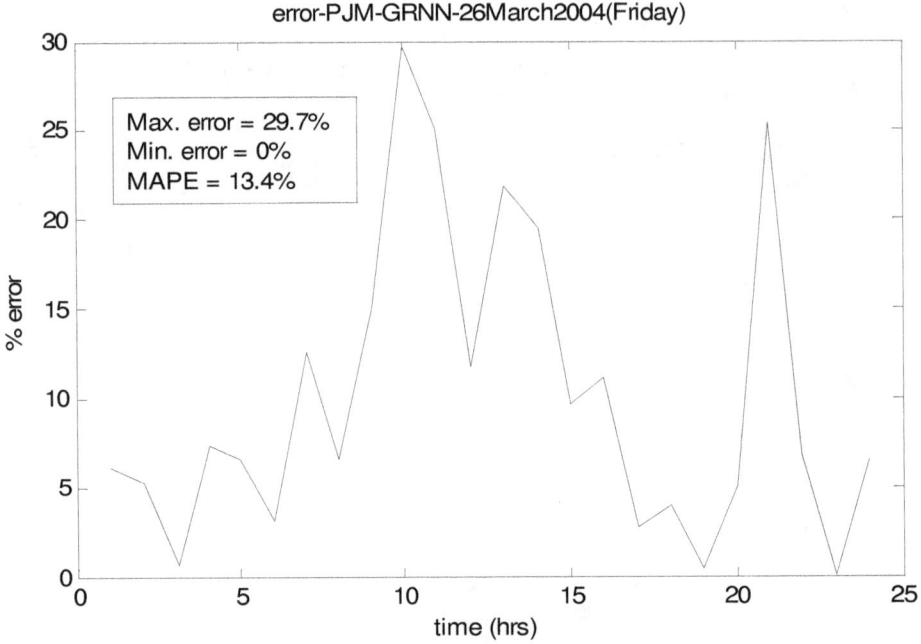

Fig. 6.21: Error in price forecast of Friday

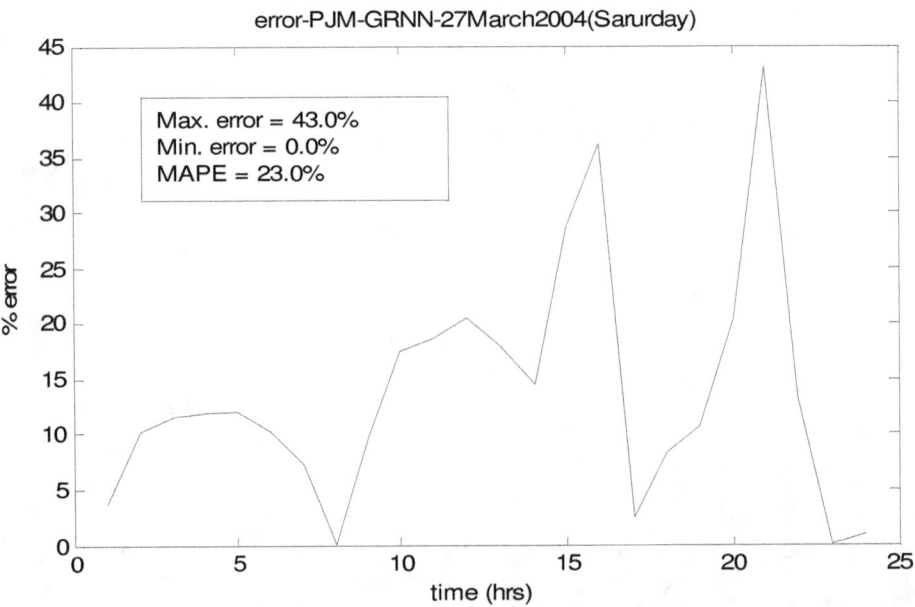

Fig. 6.22: Error in price forecast of Saturday

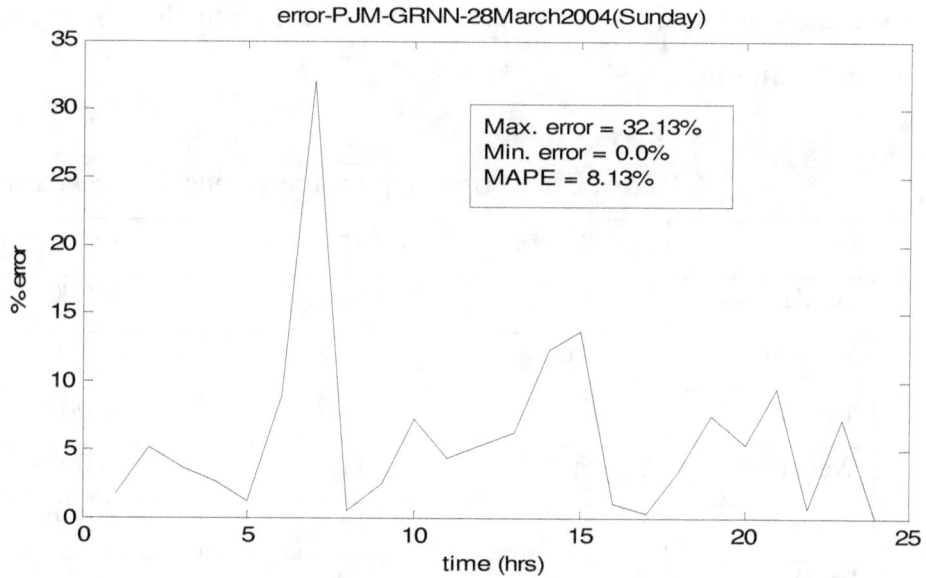

Fig. 6.23: Error in price forecast of Sunday

Forecasting result of one week (22-28 March 2004) is given in Fig. 6.24 and it is found that forecasted electricity price of one week using GRNN almost follows the actual price patterns.

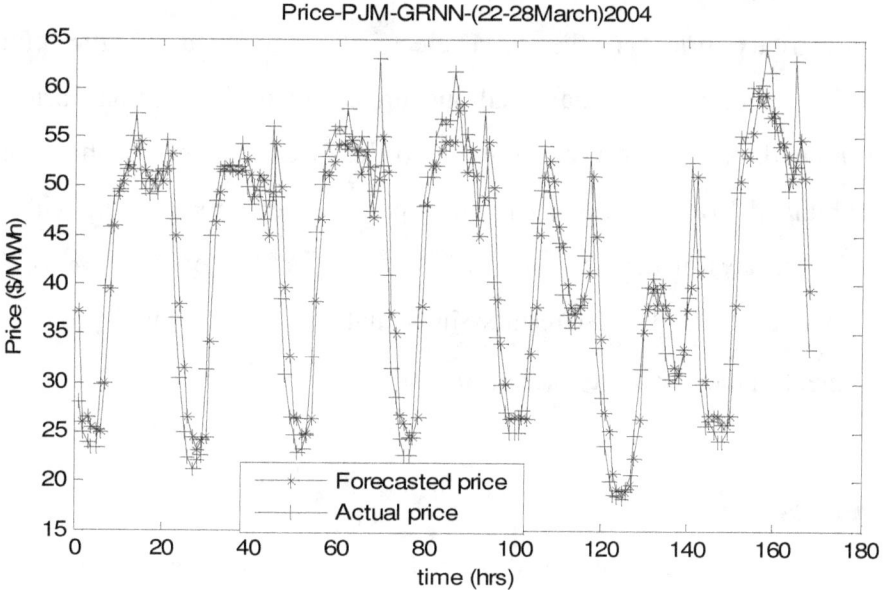

Fig. 6.24: Price forecast of one week

MAPE for all weekdays are also calculated. The MAPE and errors in forecasting the price on all weekdays are listed below in table 6.1. It is evident from the table 6.1 that MAPE on all weekdays varies from 5.54 % to 23.00 %.

Table 6.1: MAPE and errors in price forecasting for weekdays

Day	MAPE (%)	Max. Error (%)	Min. Error (%)
Sunday	8.13	32.13	0.0000
Monday	7.00	22.80	0.5600
Tuesday	7.27	17.71	0.0030
Wednesday	5.75	11.53	0.0001
Thursday	5.54	13.70	0.0001
Friday	13.40	29.70	0.0000
Saturday	23.00	43.00	0.0014

6.9 Summary

Generalized regression neural network (GRNN) has been used for the prediction of day-ahead electricity price in the PJM electricity market using the publicly available information of the previous day load and price. Historical load and price, which are the best correlated, have been used as input to the GRNN. The day-ahead price forecasting results obtained from the GRNN method provide very good result with very less error. MAPE for all days of the week are different and varies from 5.54 % to 23.00 %. MAPE values obtained from the test results show that the GRNN could provide a considerable improvement in the electricity price forecasting.

References

[1]   S. Vucetic, K. Tomsovic and Z. Obradovic, "Discovering price-load relationships in California's electricity market", IEEE Trans. on Power Systems, Vol. 16, No. 2, May 2001, pp. 280-286.

[2] T. Mount, "Market power and price volatility in restructured markets for electricity", Decision support System, Vol. 30, No. 3, 2001, pp. 311-325.

[3] S. Stoft, "Power System Economics: Designing Markets for Electricity", New York: IEEE Press, 2002.

[4] H. Y. Yamin, S. M. Shahidehpour and Z. Li, "Adaptive short-term electricity price forecasting using artificial neural networks in the restructured power markets", Electrical Power and Energy Systems, Vol. 26, 2004, pp. 571-581.

[5] Paras Mandal, Tomonobu Senjyu, Naomitsu Urasaki, Toshihisa Funabashi, and Anurag K. Srivastava, "Electricity Price Forecasting for PJM Day-Ahead Market", IEEE PES Power Systems Conference and Exposition, 2006, pp 1321-1326

[6] F. J. Nogales, J. Contreras, A. J. Conejo and R. Espinola, "Forecasting next-day electricity prices by Time Series Model", IEEE Trans. on Power Systems, Vol. 17, No. 2, August 2002, pp. 342-348.

[7] N. Amjady and M. Hemmati, "Energy price forecasting - problem and proposals for such predictions," IEEE Power and Energy Magazine, Vol. 4, No. 2, March-April 2006, pp. 20-29.

[8] PJM Web Site, http://www.pjm.com, Active March 2006.

[9] Y. Y. Hong and C. Y. Hsiao, "Locational marginal price forecasting in deregulated electricity markets using artificial intelligence", IEEE Proc. Generation, Transmission and Distribution, Vol. 149, No. 5, Sept. 2002, pp. 621-626.

[10] J. Bastian, J. Zhu, V. Banunarayanan and R. Mukerji, "Forecasting energy prices in a competitive market", IEEE Computer Application in Power, Vol. 12, No. 3, July 1999, pp. 40-45.

[11] D. W. Bunn, "Forecasting loads and prices in competitive power markets", Proceedings of the IEEE, Vol. 88, No. 2, Feb. 2000, pp. 163-169.

[12] J. Contreras, R. Espinola, F. J. Nogales and A. J. Conejo, "ARIMA models to predict next-day electricity prices", IEEE Trans. on Power Systems, Vol. 18, No. 3, Aug. 2003, pp. 1014-1020.

[13] P. Mandal, T. Senjyu and T. Funabashi, "Neural networks approach to forecast several hour ahead electricity prices and loads in deregulated market", Energy Conversion and Management, Vol. 47, No. 15-16, September 2006, pp. 2128-2142.

[14] B. R. Szkuta, L. A. Sanabria and T. S. Dillon, "Electricity price short term forecasting using ANN", IEEE Trans. on Power Systems, Vol. 14, No. 3, Aug. 1999, pp. 851-857.

[15] C. P. Rodriguez and G. J. Anders, "Energy price forecasting in the Ontario competitive power system market", IEEE Trans. on Power Systems, Vol. 19, No. 3, Feb. 2004, pp. 366-374.

[16] A.J. Conejo, J. Contrearas, R. Espinola and M.A. Plazas, "Forecasting Electricity Prices for a day ahead pool based electric energy market", International Journal of Forecasting, Vol. 21, 2005, pp. 435–462.

[17] A. M. Gonzalez, A. M. S. Roque and J. Gargia-Gonzalez, "Modeling and forecasting electricity prices with input/output hidden Markov models", IEEE Trans. on Power Systems, Vol. 20, No. 1, 2005, pp. 13-24.

[18] R. Gareta , L. M. Romeo and A. Gil, "Forecasting of electricity prices with neural networks", Energy Conversion and Management Vol. 47, No. 13-14, 2003, pp. 1770-1778.

[19] H. S. Hippert, C. E. Pedreira and R. C. Souza, "Neural networks for short-term load forecasting: A review and evaluation", IEEE Trans. on Power Systems, Vol. 16, No. 1, 2001, pp. 44-55.

[20] T. Senjyu, P. Mandal, K. Uezato and T. Funabashi, "Next day load curve forecasting using hybrid correction method", IEEE Trans. on Power Systems, Vol. 20, No. 1, 2005, pp. 102-109.

[21] T. Senjyu, P. Mandal, K. Uezato and T. Funabashi, "Next day load curve forecasting using recurrent neural network structure", IEEE Proc. Generation, Transmission and Distribution, Vol. 151, No. 3, May 2004, pp. 388-394.

[22] E. NI and P. B. Luh, "Forecasting power market clearing price and its discrete PDF using a bayesian-based classification method", IEEE PES Winter Meeting, Vol. 28, Colombus, OH, USA, 2001.

[23] F. Gao, X. Guan, X. R. Cao and A. Papalexopoulos, "Forecasting power market clearing price and quantity using a neural network", IEEE PES Summer Meeting, Seattle, WA, 2000.

[24] Paras Mandal, Tomonobu Senjyu, Naomitsu Urasaki, Toshihisa Funabashi, Anurag K. Srivastava, "A Novel Approach to Forecast Electricity Price for PJM Using Neural Network and Similar Days Method" IEEE TRANSACTIONS ON POWER SYSTEMS, VOL. 22, NO. 4, NOVEMBER 2007, pp 2058-2065

[25] M. Shahidehpour and M. Alomoush, "Restructured Electrical Power Systems: Operation, Trading and Volatility", New York: Marcel Dekker, 2001.

[26] Y. S. Son, R. Baldick, K.Lee and S. Siddiqi, "Short-term electricity market auction game analysis: Uniform and pay-as-bid pricing", IEEE Trans. on Power Systems, Vol. 20, No. 2, May 2005, pp. 1035–1042.

[27] M. Shahidehpour, H. Yamin and Z. Li, "Market Operations in Electric Power Systems: Forecasting, Scheduling, and Risk Management", New York: Wiley, Apr. 2002.

[28] J. Contreras, R. Espínola, F. J. Nogales, and A. J. Conejo, "ARIMA Models to predict next

day electricity prices," IEEE Trans. on Power Systems, vol. 18, no. 3, pp. 1014–1020, August 2003

[29] F. J. Nogales, J. Contreras, A. J. Conejo, and R. Espínola, "Forecasting Next-Day Electricity Prices by Time Series Models," IEEE Trans. on Power Systems, vol. 17, no. 2, pp. 342–348, May 2002

[30] A.J. Conejo, J. Contrearas, R. Espinola, and M.A. Plazas, "Forecasting Electricity Prices for a day ahead pool based electric energy market," International Journal of Forecasting, vol. 21, pp. 435–462, 2005.

[31] E. Weiss, "Forecasting commodity prices using ARIMA," Technical Analysis of Stocks & Commodities, vol. 18, no. 1, pp. 18–19, 2000.

[32] C. Morana, "A semiparametric approach to short-term oil price forecasting," Energy Economics, vol. 23, no. 3, pp. 325–338, May 2001.

[33] W. K. Buchananan, P. Hodges, and J. Theis, "Which way the natural gas price: An attempt to predict the direction of natural gas spot price movements using trader positions," Energy Economics, vol. 23, no. 3, pp. 279–293, May 2001.

[34] G. Gross and F. D. Galiana, "Short-Term load forecasting," Proc. IEEE, vol. 75, no. 12, pp. 1558–1573, Dec. 1987.

[35] O. B. Fosso, A. Gjelsvik, A. Haugstad, M. Birger, and I. Wangensteen, "Generation scheduling in a deregulated system. The norwegian case," IEEE Trans. Power Systems, vol. 14, no. 1, pp. 75–81, Feb. 1999.

[36] E. Ni and P. B. Luh, "Forecasting Power Market Clearing Price and Its Discrete PDF Using a Bayesian-based Classification Method," IEEE Power Engineering Society Winter Meeting, vol. 3, pp. 1518–1523, Columbus, OH, 2001.

[37] M. Zhou, Z. Yan, Y. Ni and G. Li, "An ARIMA approach to forecasting electricity price with accuracy improvement by predicted errors," in Proc. IEEE Power Engineering Society General Meeting, pp. 233–238, June 2004.

[38] H. Zareipour, K. Bhattacharya and C.A. Ca˜nizares, " Forecasting the Hourly Ontario Energy Price by Multivariate Adaptive Regression Splines", in Proc. the IEEE Power Engineering Society General Meeting, 18–22 June, 2006, Montreal, Quebec, Canada.

[39] R.C. Garcia, J. Contreras, M. van Akkeren, and J.B.C. Garcia, "A GARCH Forecasting Model to Predict Day-Ahead Electricity Prices," IEEE Trans. on Power Systems, vol. 20, no. 2, pp. 867 – 874, May 2005.

[40] Parviz Doulai and Warren Cahill, "Short Term Price Forecasting in Electric Energy market", International Power Engineering Conf. (IPEC 2001), May 2001, pp. 749–754.

[41] Paras Mandal, Tomonobu Senjyu, Katsumi Uezato, and Toshihisa Funabashi, "Several-Hours-Ahead Electricity Price and Load Forecasting Using Neural Networks ", IEEE Power Engineering Society General Meeting, vol. 3, pp. 2146–2153, 12-16 June 2005.

[42] Alireza Sedaghati, "Using Neural Network to Forecast Price in Competitive Power Market", International Conf. on Control, Automation & System (ICCAS), June 2–5, 2005, Korea.

[43] L. Zhang, P. B. Luh, and K. Kasiviswanathan, "Energy Clearing Price Prediction and Confidence Interval Estimation With Cascaded Neural Networks," IEEE Trans. on Power Systems, vol. 18, no. 1, pp. 99–105, Feb. 2003.

[44] J.-J. Guo, and P.B. Luh, "Improving Market Clearing Price Prediction by Using a Committee Machine of Neural Networks," IEEE Trans. on Power Systems, vol. 19, no. 4, pp. 1867–1876, Nov. 2004.

[45] B.R. Szkuta, L.A. Sanabria and T.S. Dillon, "Forecasting Power Market Clearing Prices and Quantity Using Neural Network Method", IEEE Power Engineering Society Summer Meeting, pp. 2183–2188, 2000.

[46] F. Gao, X. Guan, X. Cao, and A. Papalexopoulos, "Forecasting Power Market Clearing Price and Quantity Using a Neural Network Method," in Proc. of Power Engineering Summer Meeting, Seattle, WA, July 2000, vol. 4, pp. 2183–2188.

[47] T. Niimura and T. Nakashima, "Deregulated electricity market data representation by fuzzy regression models," IEEE Trans. on Systems, Man and Sybernetics–Part c: Applications and Reviews, vol. 31, no. 3, pp. 320–326, Aug. 2001.

[48] L. Hongjie, W. Xiugeng, Z. Weicun, and X. Guohua, "Market clearing price forecasting based on dynamic fuzzy system," in Proc. PowerCon 2002, International Conf. on Power System Technology, vol. 2, pp. 890–896, 13–17 Oct. 2002.

[49] L. Zhang and P.B. Luh, "Neural network-based market clearing price prediction and confidence interval estimation with an improved extended Kalman filter method," IEEE Trans. on Power Systems, vol. 20, no. 1, pp. 59–66, Feb. 2005.

[50] Dong-Xiao Niu, Hui-Qing Wang and Zhi-Hong Gu, "Short-Term Load Forecasting using general regression neural network", Proceedings of the Fourth International Conference on Machine Learning and Cybernetics, Guangzhou, Vol. 7, 18-21 August 2005, pp. 4076-4082.

[51] Menniti Daniele, Scordino Nadia, Sorrentino Nicola, "A Novel Approach to Forecast Day-Ahead Electricity Prices by Means of Neural Networks Using Groups of Similar Hours", International Review of Electrical Engineering, 2012.

CHAPTER 7

Spinning Reserve (SR) Forecasting

7.1   Introduction

Based on the trading protocols, competitive electricity markets may include separate energy market, transmission market and ancillary services (AS) market. Ancillary services are necessary to support the transmission of power from sellers to buyers by maintaining the reliable and stable operation of the interconnected power system. The reliable operation of power system requires generation reserves to be available in order to cover generation and transmission contingencies. The independent system operator (ISO) in many countries has the overall responsibility of providing and procuring the required AS through competitive bidding process on behalf of market participants to operate the grid in a secure and reliable manner. The AS procured competitively, on a daily basis, by California ISO (CAISO) are: regulation reserves, spinning reserve and non-spinning reserve. The spinning and non-spinning reserves are together called operating reserve (OR).

The spinning reserve (SR) is the fastest-responding contingency reserve and thus, the most critical for maintaining power system reliability following a major contingency such as the unplanned outage of a large generation facility or critical transmission line. As the SR requirements have significant impacts on the energy market [1], the accurate forecasting of day-ahead SR requirements helps the market participants (the scheduling coordinators) to develop an optimal bidding strategy that would maximize their profits in the market. An accurate short-term predication of day-ahead AS requirement helps the ISO to make effective and timely decisions in managing the compliance and reliability of the power system.

The CAISO prepares and publishes an hourly forecast of SR requirements for entire CAISO system based on the CAISO load forecast. The forecast accuracy of SR requirements depends on accuracy of load forecast. The accuracy in SR requirements forecast may be improved if forecasting is done on the basis of historical SR requirements time series data itself. This also helps to take care of additional source of error for load forecast. SR requirements data exhibits non-linear and non-stationary characteristics, therefore, the time series models like auto-regressive integrated moving average (ARIMA), transfer function (TF), etc., which are having limited ability to capture such type characteristics, may not accurately predict hourly SR requirements. Generalized regression neural network (GRNN) has been used for short-term load and price forecasting in electricity market and it has been found that GRNN performed better than other load and price forecasting methods [2], [3]. A generalized regression neural network (GRNN) is used to forecast day-ahead spinning reserve in California market. The results shown in this chapter indicates that GRNN is able to forecast SR for winter as well as summer seasons with less error.

7.2 Ancillary Services in California Electricity Market

In California electricity market, the energy market and the ancillary service market are managed separately by two different entities, i.e. the power exchange (PX) and the ISO, respectively. CAISO procures AS requirements on an hourly basis in both the day-ahead and hour-ahead markets in a sequential manner to ensure compliance with western electricity coordinating council's (WECC) minimum operating reliability criteria (MORC) [4], [5] .

7.2.1 Regulation Reserves

Regulation reserves are the generating resources that are running and synchronized with the CAISO controlled grid so that the operating levels can be increased (incremented) or decreased (decremented) instantly through automatic generation control (AGC). It is used to maintain real-time power balance to maintain frequency of the system.

7.2.2 *Spinning Reserves*

Spinning reserves are the generating resources that are running (spinning) with additional capacity, capable of ramping over a specified range within 10 minutes and running for at least two hours. It is needed to maintain system frequency stability during emergency operating conditions and unforeseen load swings.

7.2.3 *Non-Spinning Reserves*

Non-spinning reserves are the generating resources that are available but not running. These are capable of being synchronized to the grid and ramped to a specified level within 10 minutes and then run for at least two hours.

The CAISO operates a separate market for each of the AS. Scheduling coordinators (SCs) that wish to provide AS to CAISO may either bid for AS bids or self-provide AS. Submitted AS bids must contain a capacity component and an energy component. The selection of AS suppliers is based on the capacity bids and the required AS amount by the CAISO. It has been observed that acceptance of self-provided AS occurs prior to AS bid evaluation in the relevant market. It reduces the AS requirements that need to be met by AS bids within the same AS region and also reduces the AS obligation for the SC who themselves are self-providing the AS, in the AS cost allocation [4]. Although firms submit bids in AS markets simultaneously, the markets clear sequentially and separately in the following order: regulation, spinning reserve and non-spinning reserve. The capacity accepted by the CAISO in one of these markets is not passed on to the other markets. However, any losing bids in one market may be passed onto the next market on owner's request. Thereafter, the CAISO uses the energy bid prices of the dispatch units to provide real-time energy. Thus, besides bidding in the energy market, each supplier may have an interest in developing bidding strategies in AS market.

7.3 Architecture and Input Selection for ANN

Generalized regression neural network (GRNN) has been used for spinning reserve forecasting of California electricity market. The detailed architecture of GRNN has been discussed in Chapter 3. The publicly available day-ahead hourly SR requirements of CAISO controlled grid [6], for the year 2007, of the 48 days previous to the day for which SR requirements are to be forecasted have been considered. The next hour forecasts are performed for each hour of the day. The concatenation of 48 days training windows, for a particular day, is shifted one day ahead, and the forecasts for the next 24 hours are evaluated [7]. To demonstrate the performance of GRNN model, the last week of winter and summer seasons have been considered as test weeks. The first test week corresponds to the last week of February 2007 (19-25 February 2007) and second one corresponds to the last week of August 2007 (from Aug. 24th to Aug. 30th). The hourly data used to forecast the winter and summer test weeks are from January 2nd to Feb. 18th, 2007 and from July 2nd to August 23rd, 2007, respectively.

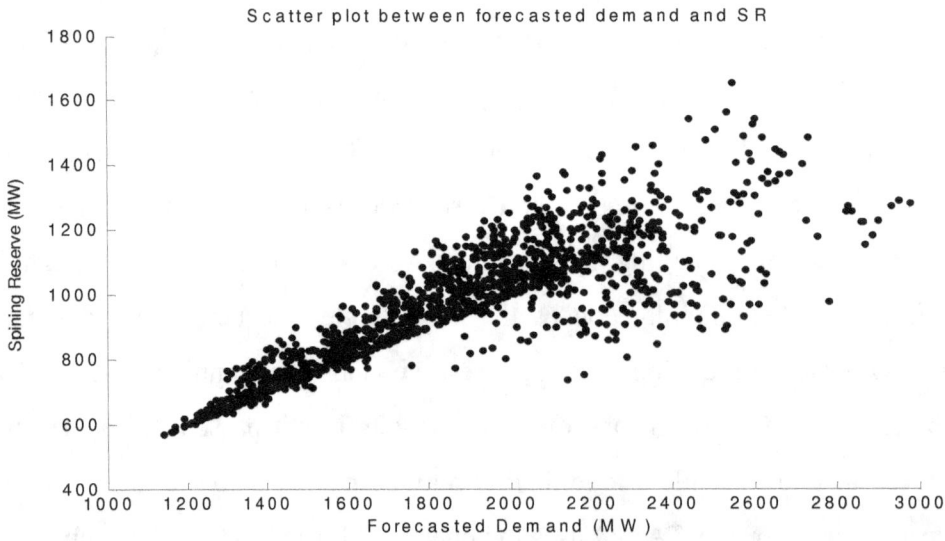

Fig. 7.1: CASIO forecasted demand versus spinning reserve requirement

CAISO publishes an hourly forecast of SR requirements for the entire CAISO system based on the CAISO demand forecast. The correlation between SR requirements and forecasted demand is presented in fig. 7.1. This clearly shows that SR requirements are highly correlated with the forecasted demand and therefore, it is used as a variable to predict the SR requirements of the CAISO controlled grid. From fig. 7.2, it is clear that the day-ahead hourly AS requirement series exhibits multiple seasonal patterns corresponding to daily and weekly seasonality.

Fig. 7.3 shows the auto correlation function (ACF) for ST time series of Fig. 7.2, with a lag time of up to 336 hour, showing daily periodicities. The ACF curve indicates that spinning reserve at the hour of prediction, i.e. SR_h, and lags up to the spinning reserve measured 3 hours earlier than the hour of prediction, i.e. SR_{h-3}, are highly correlated. Therefore, information of three previous hours can be used to model the trend of the signal. Fig. 7.4 shows the cross correlation function (XCF) between hourly SR requirements and forecasted demand (FD) by the CAISO. The XCF is useful to decide the number of lagged values of CAISO forecasted demand to predict the SR requirements.

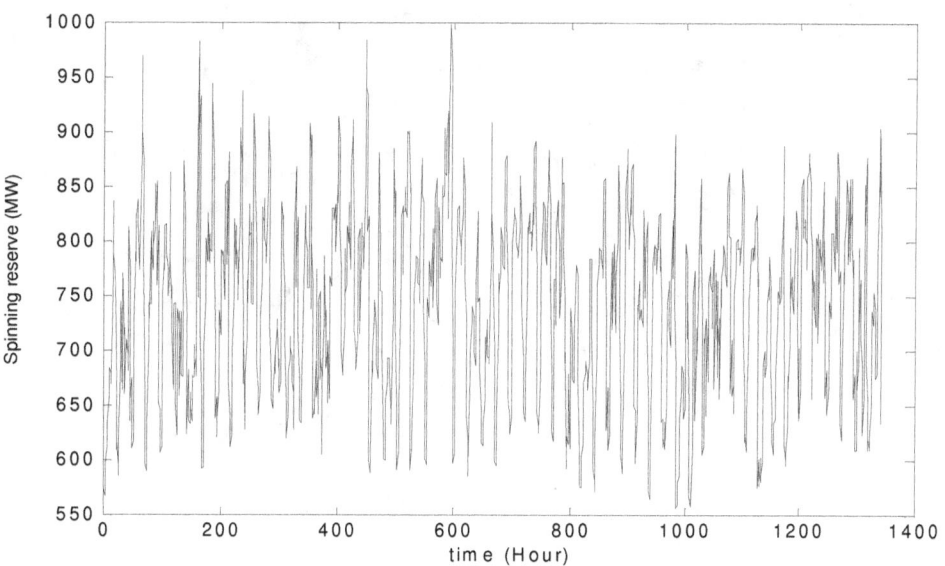

Fig. 7.2: Hourly spinning reserve (SR) data

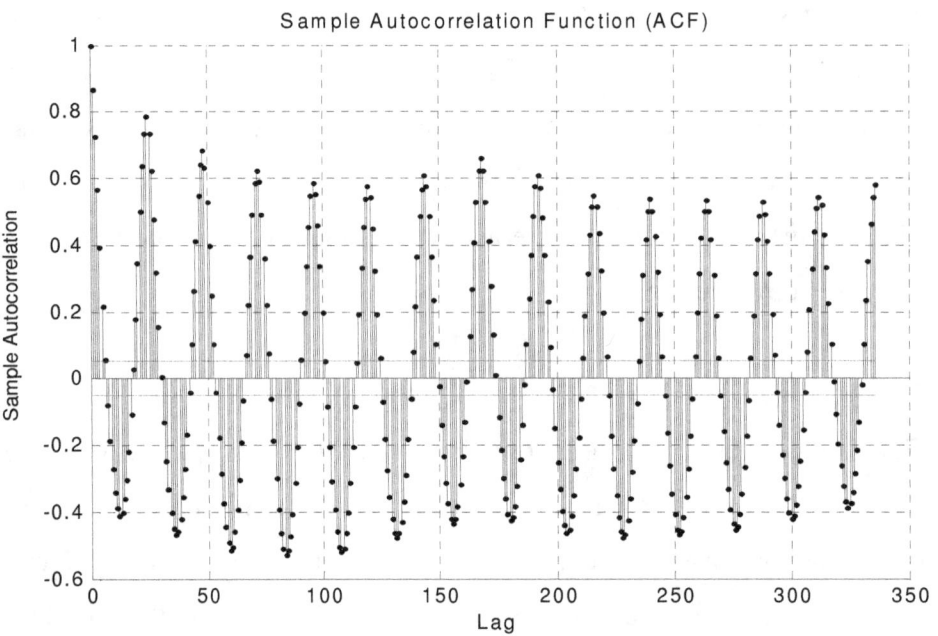

Fig. 7.3: Auto-correlation plot of SR time series

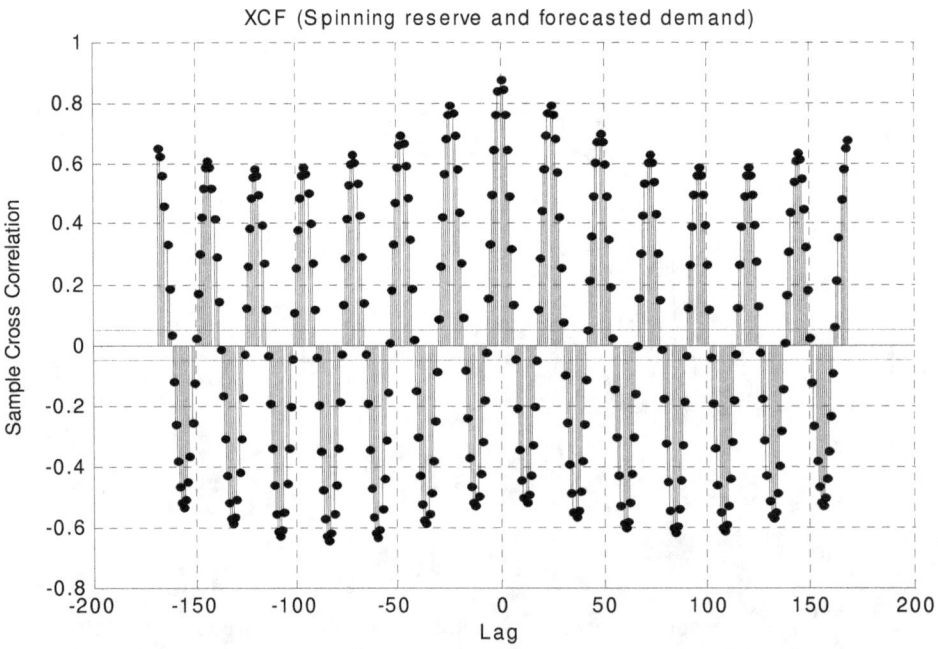

Fig 7.4: Cross correlation plot between forecasted demand and SR

7.4   ANN Training Data and Algorithm

With consideration of all the above-discussed factors, a set of input variables has been used to forecast the spinning reserve SR_h at hour h. The total number of variables are 24 consisting of 14 SR and 10 forecasted demand data. As mentioned earlier, the historical hourly spinning reserve data of the last 48 days prior to the day whose SR requirements is to be predicted have been considered to build the forecasting model.

7.5 Simulation Results with GRNN

The developed ANN models are trained using set of input/ output data. All the data are normalized before training. Due to its special features as explained in the previous sections, the algorithm results in a very fast training and the error is significantly reduced to very low value. Then performance of the developed GRNN model for forecasting SR requirement is tested using data sets that are not included in the training set.

The performance of the forecasting models is calculated in terms of daily mean absolute percentage error (MAPE) given by (7.1) where L_A is the actual SR, L_F is the forecasted SR, N is the number of hours.

$$MAPE(\%) = \frac{1}{N} \sum_{i=1}^{N} \frac{\left| L_F^i - L_A^i \right|}{L_A^i} * 100 \qquad (7.1)$$

The smaller values of daily MAPE means the predicted values are closer to the actual values.

Fig. 7.5, Fig. 7.7, Fig. 7.9, Fig. 7.11, Fig. 7.13, Fig. 7.15 and Fig. 7.17 show actual and forecasted SR of all days of the week (19-25 February 2007) of winter season. Fig. 7.6,

Fig. 7.8, Fig. 7.10, Fig. 7.12, Fig. 7.14, Fig. 7.16 and Fig. 7.18 show the error in SR forecasting of all days of the week (19-25 February 2007) with GRNN.

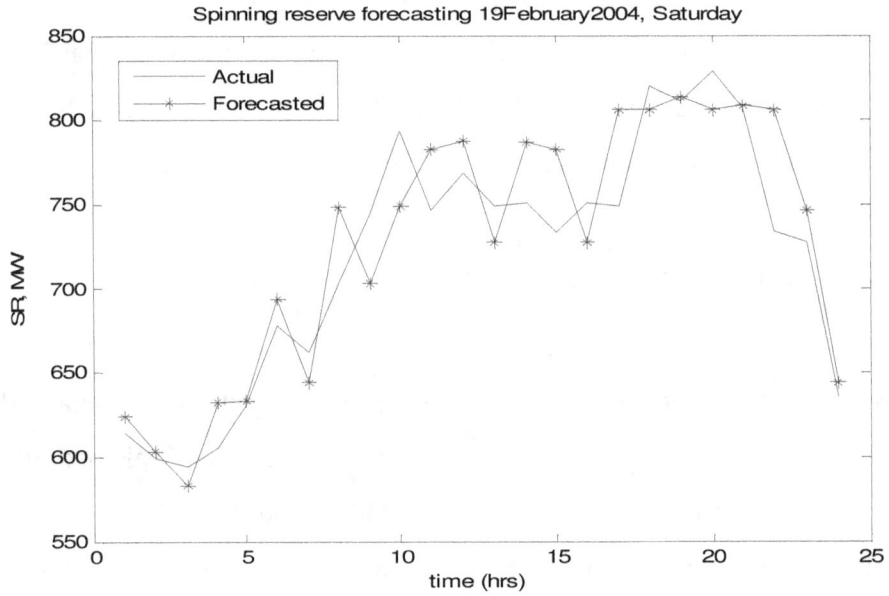

Fig 7.5: SR forecast of Saturday with modified GRNN (winter)

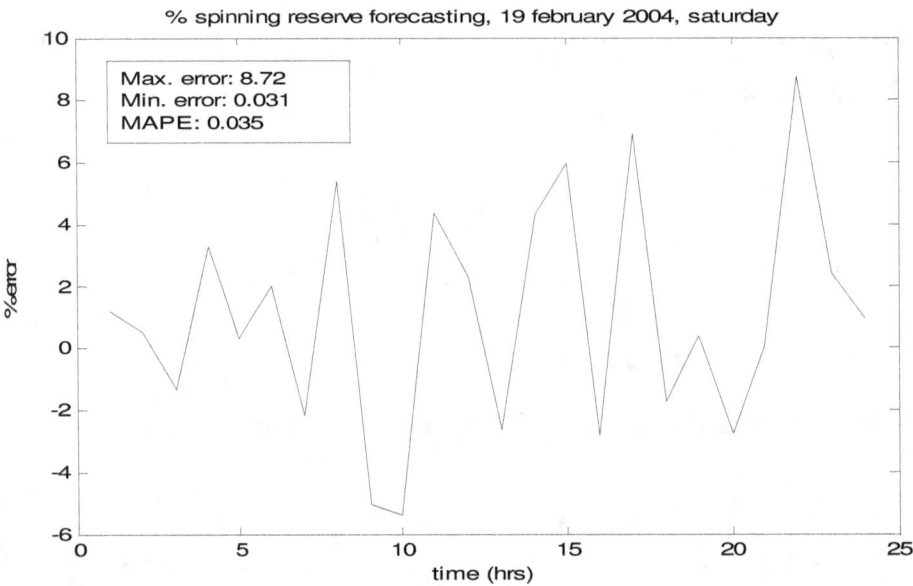

Fig 7.6: Error in SR forecast of Saturday (winter)

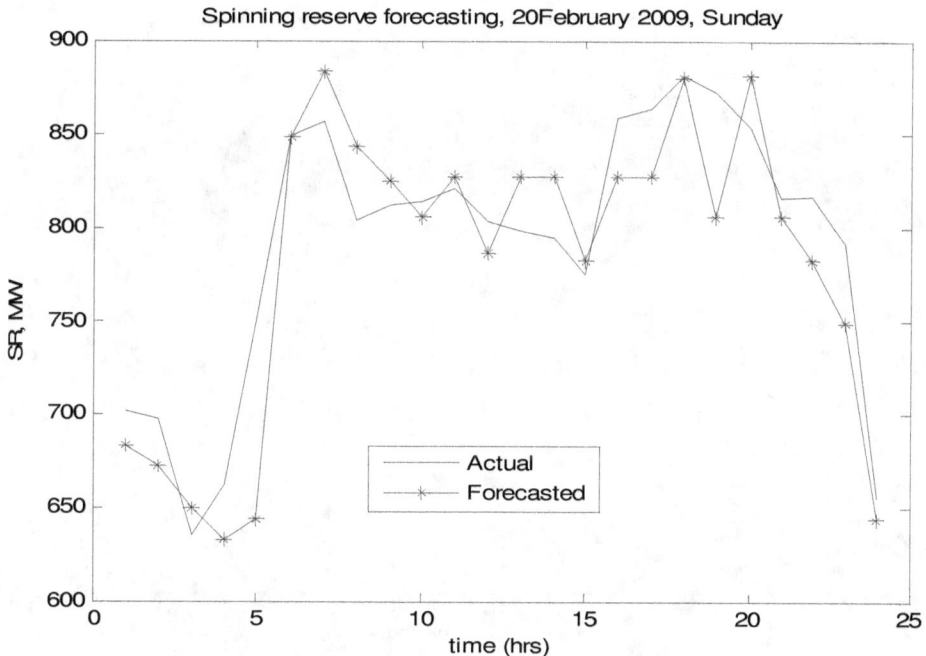

Fig 7.7: SR forecast of Sunday (winter)

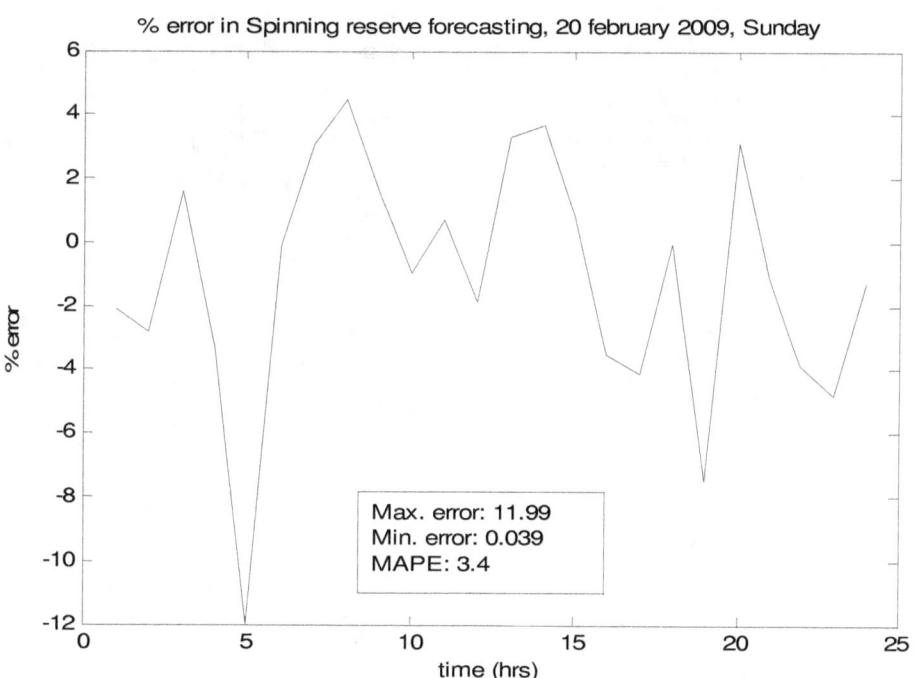

Fig 7.8: Error in SR forecast of Sunday (winter)

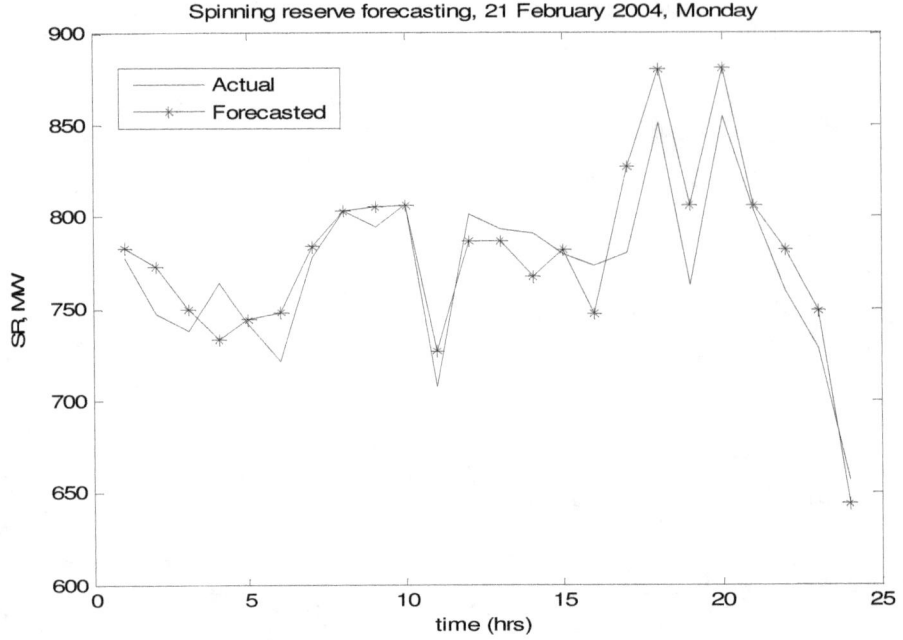

Fig 7.9: SR forecast of Monday (winter)

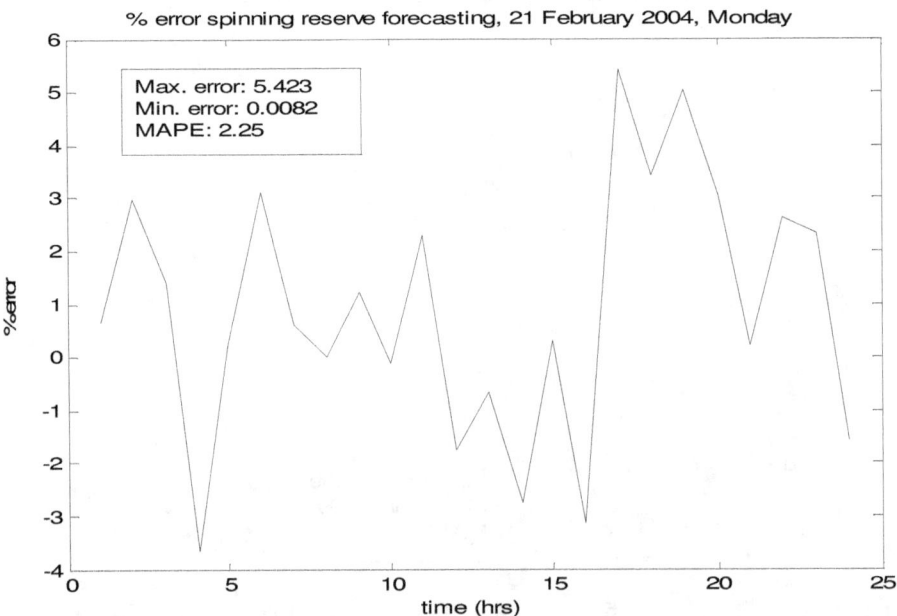

Fig 7.10: Error in SR forecast of Monday (winter)

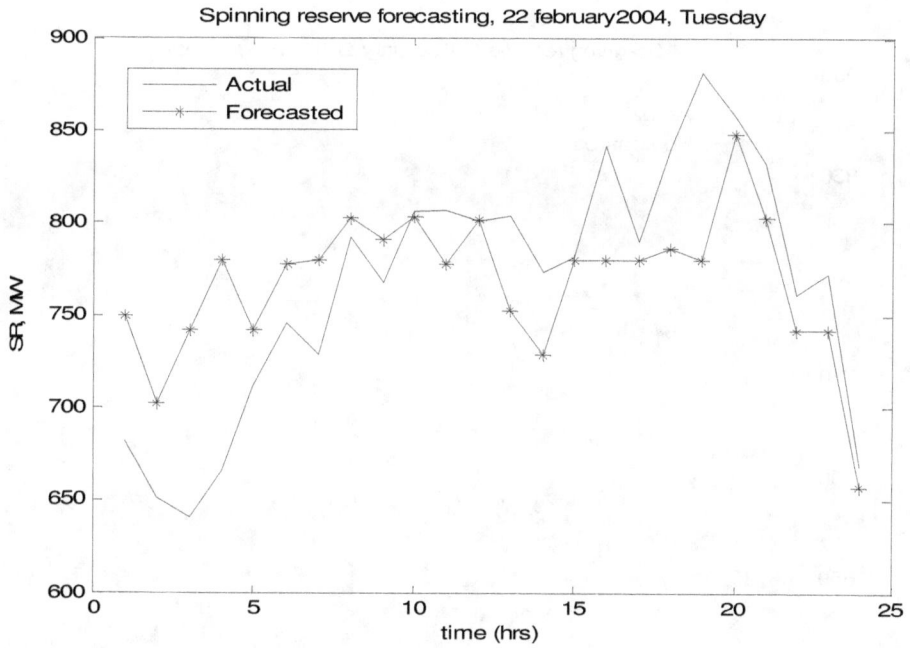

Fig 7.11: SR forecast of Tuesday (winter)

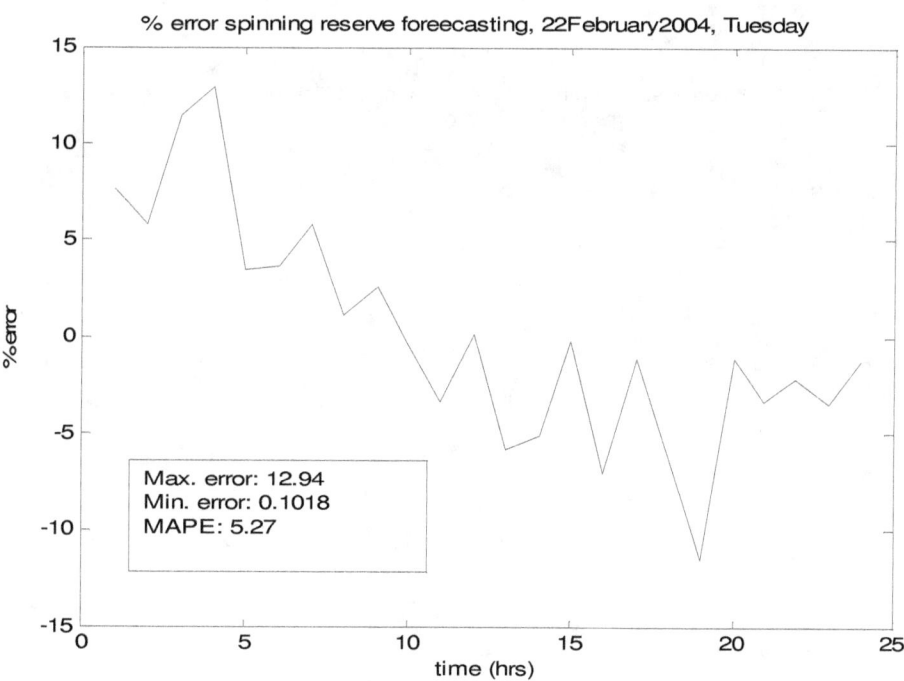

Fig 7.12: Error in SR forecast of Tuesday (winter)

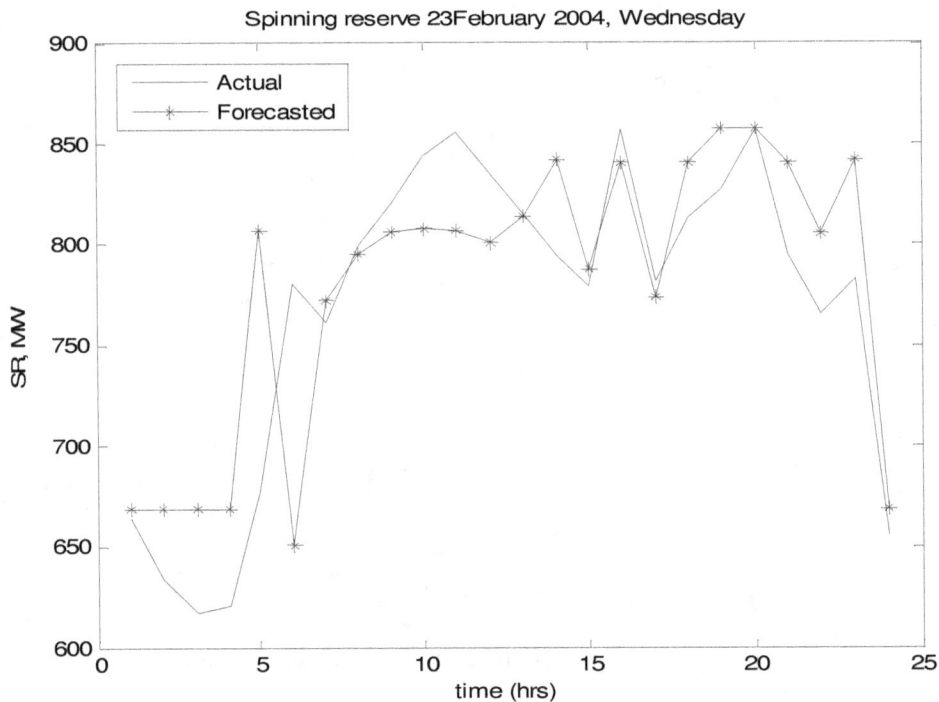

Fig 7.13: SR forecast of Wednesday (winter)

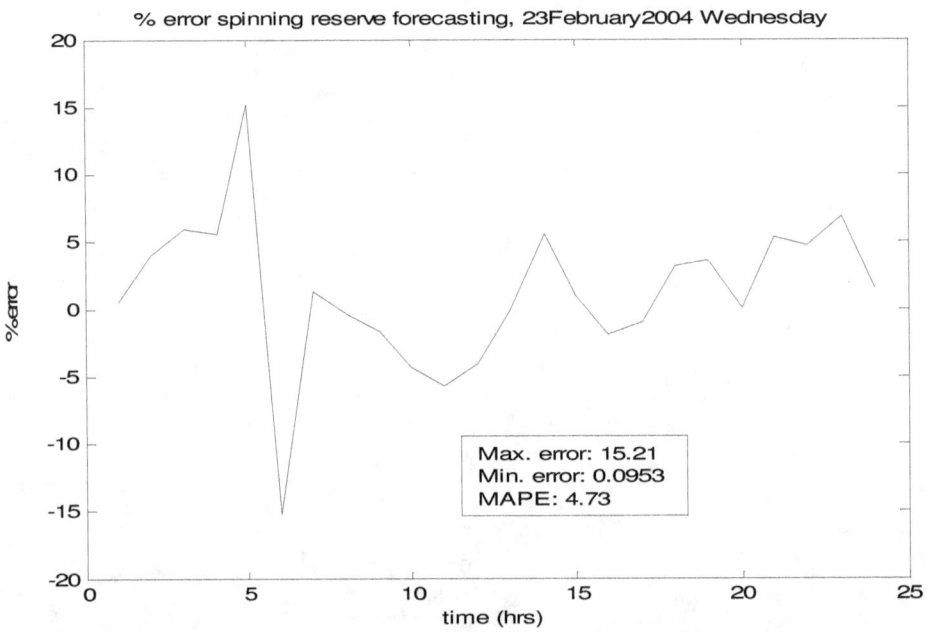

Fig 7.14: Error in SR forecast of Wednesday (winter)

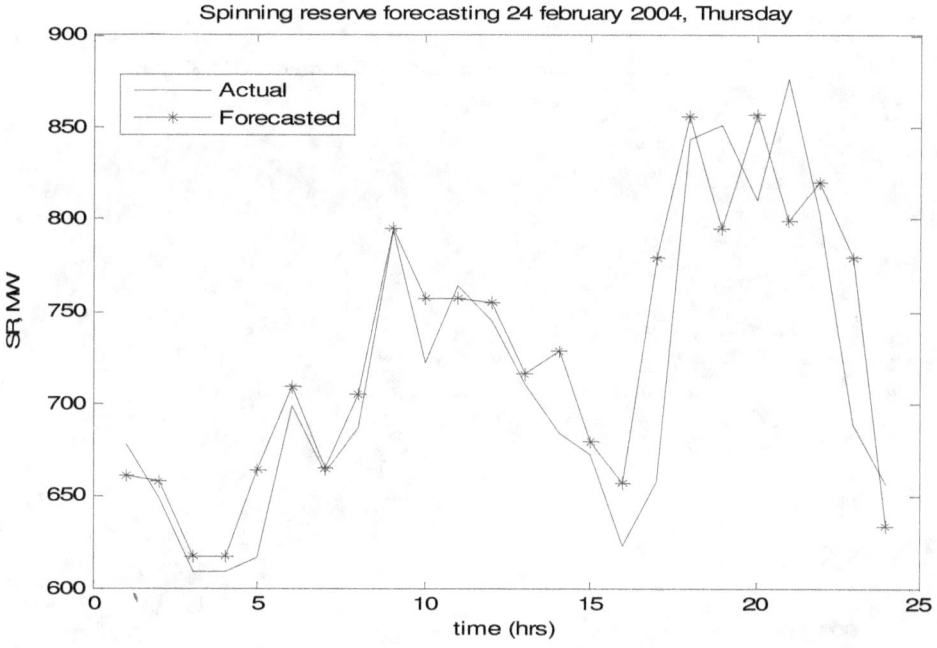

Fig 7.15: SR forecast of Thursday (winter)

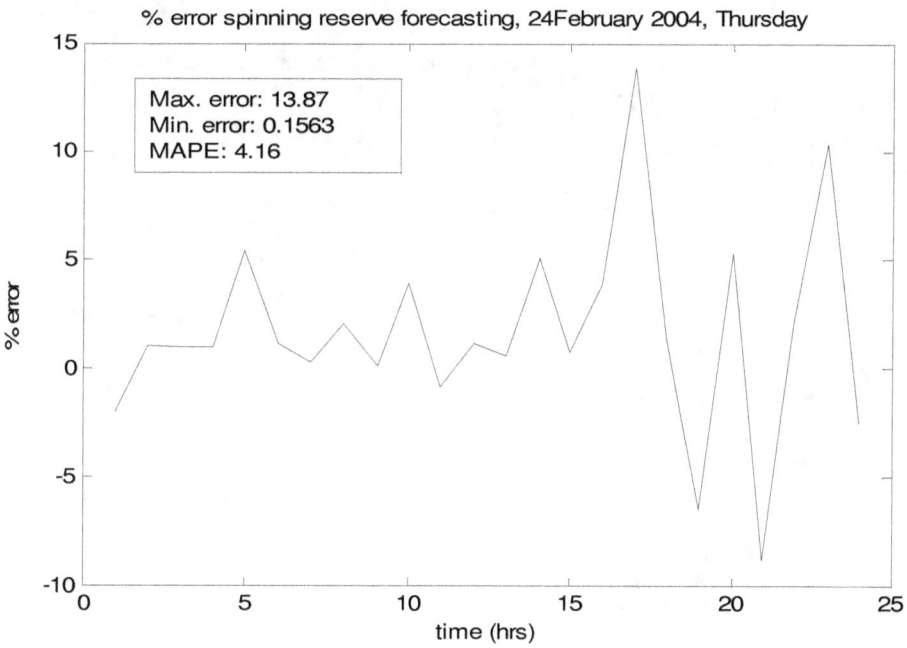

Fig 7.16: Error in SR forecast of Thursday (winter)

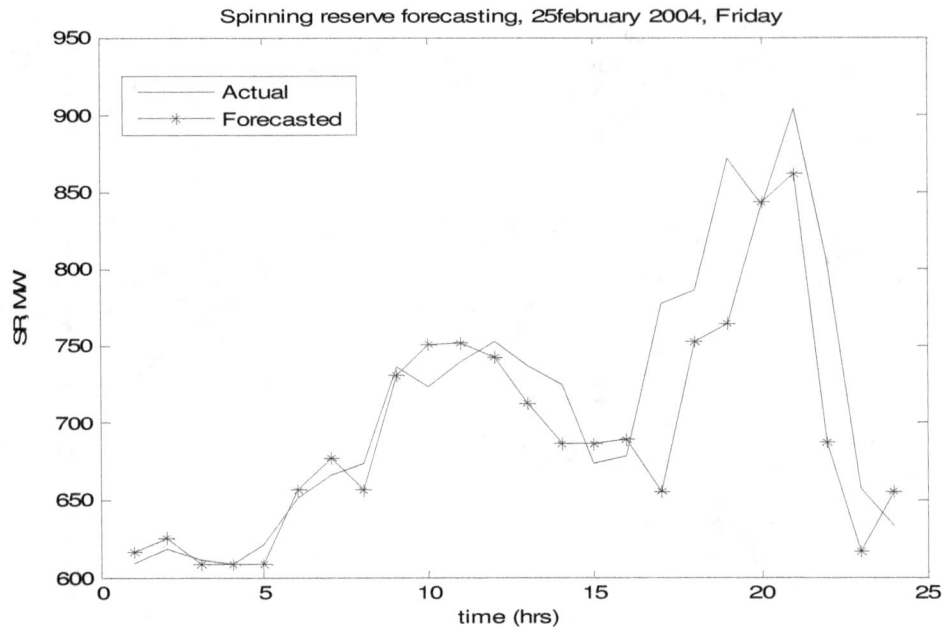

Fig 7.17: SR forecast of Friday (winter)

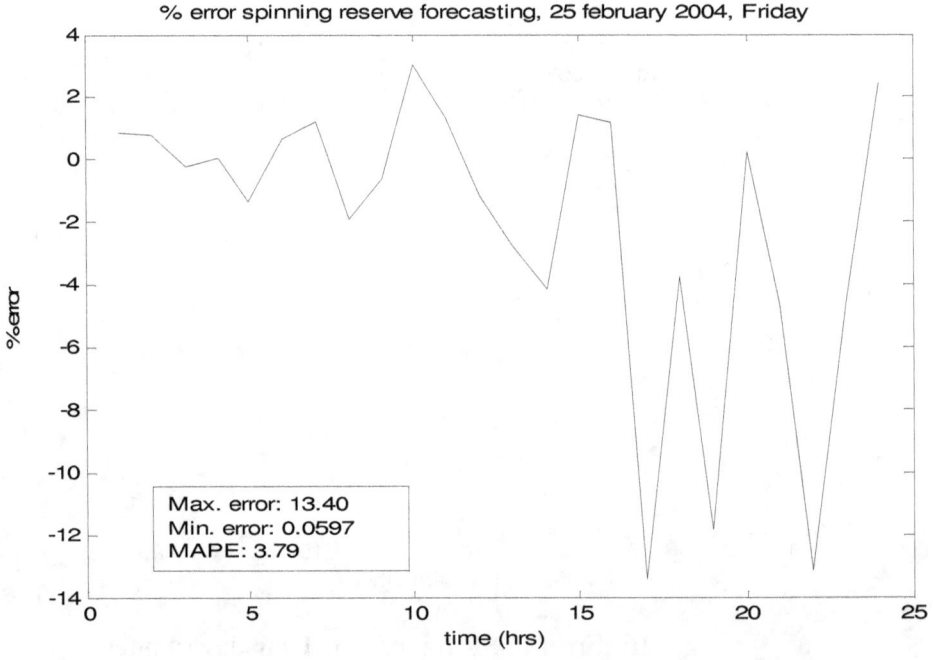

Fig 7.18: Error in SR forecast of Friday (winter)

The maximum error in SR forecasting during winter season, with GRNN varies from 5.4% to 15.21%, whereas minimum error varies from 0.008% to 0.15%.

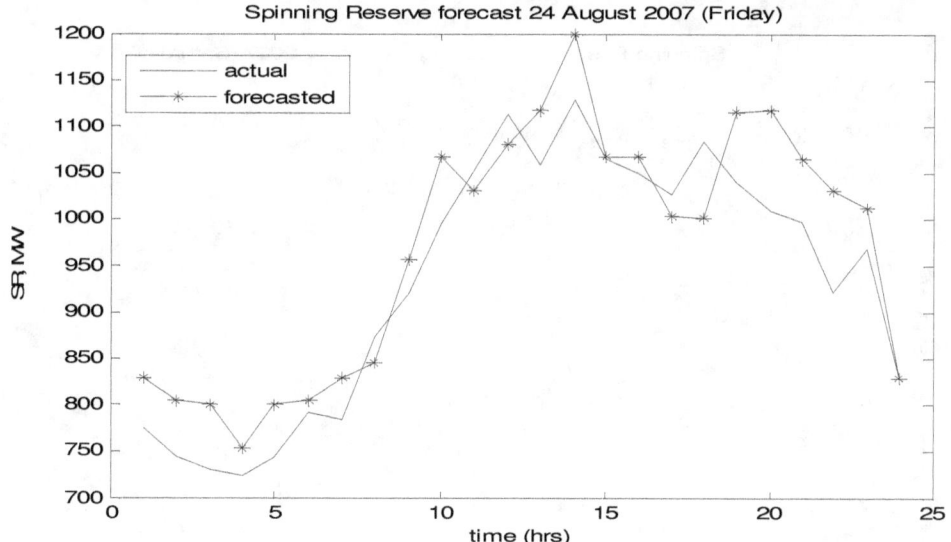

Fig 7.19: SR forecast of Friday (summer)

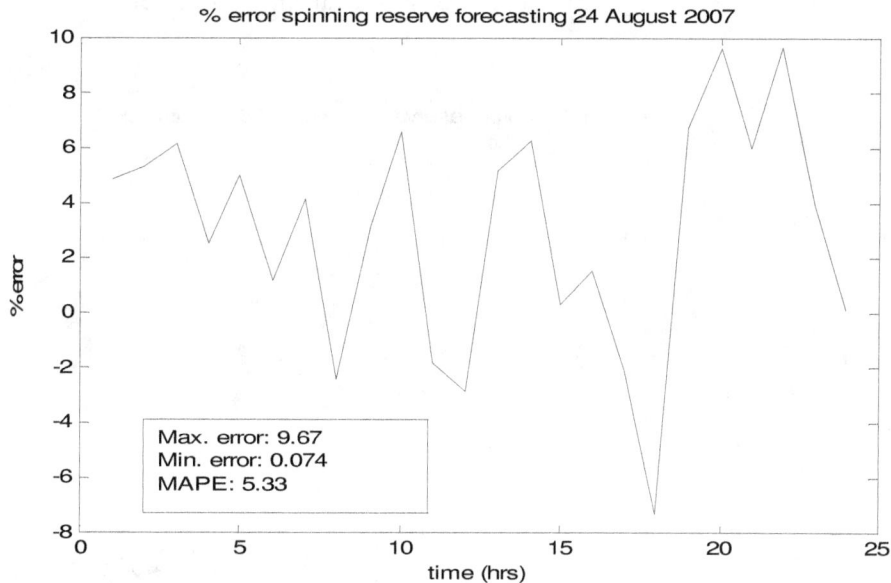

Fig 7.20: Error in SR forecast of Friday (summer)

Fig. 7.19, Fig. 7.21, Fig. 7.23, Fig. 7.25, Fig. 7.27, Fig. 7.29 and Fig. 7.31 show actual and forecasted SR of all days of the week (24-30 August 2007) of summer season. Fig. 7.20, Fig. 7.22, Fig. 7.24, Fig. 7.26, Fig. 7.28, Fig. 7.30 and Fig. 7.32 show the error in SR forecasting of all days of the week (24-30 August 2007) of summer season with GRNN.

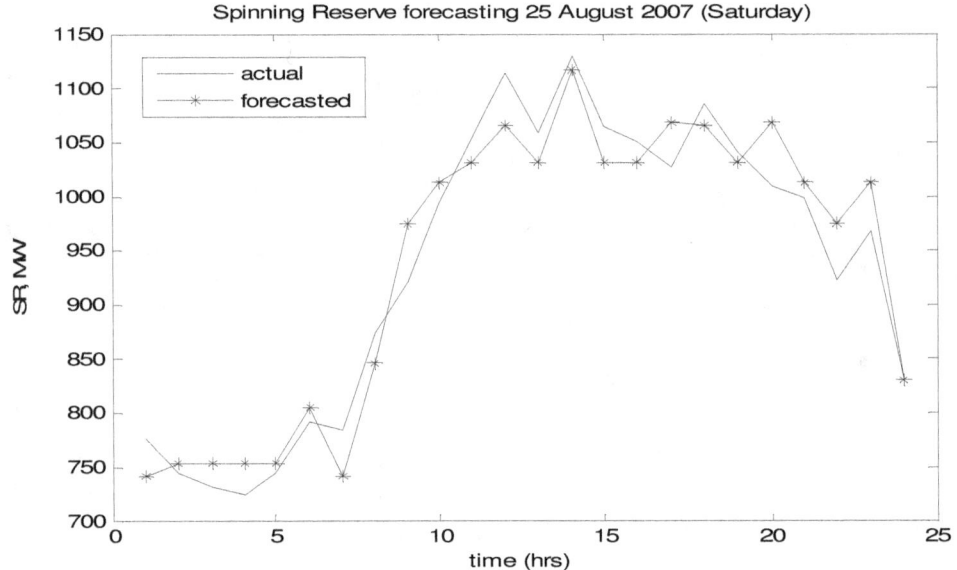

Fig 7.21: SR forecast of Saturday (summer)

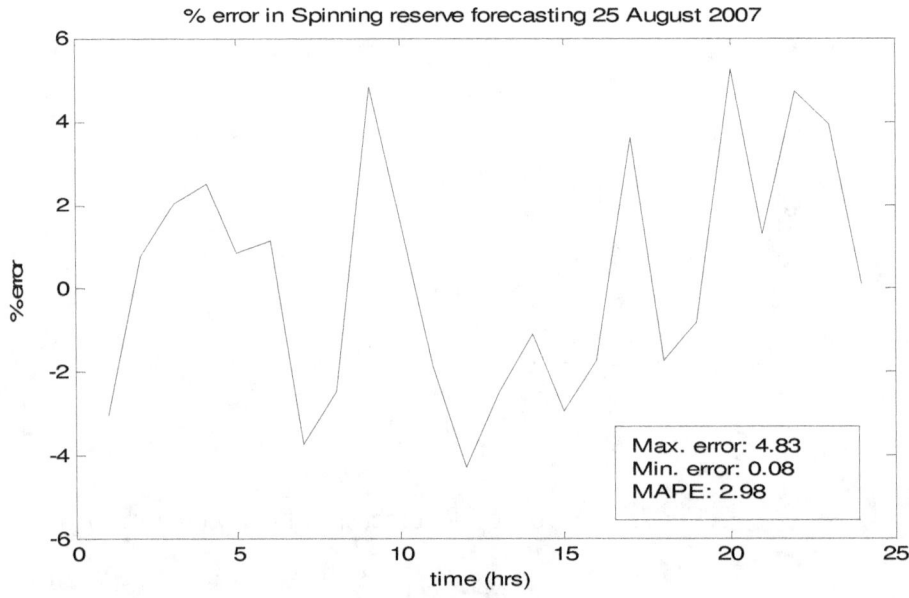

Fig 7.22: Error in SR forecast of Saturday (summer)

During summer season, maximum error in SR forecasting with GRNN varies from 4.83% to 17.55%, whereas minimum error varies from 0.003% to 1.33%. It is also evident from the figures that forecasted SR requirement curve almost follows the actual SR requirement curve on Saturday, Sunday, Monday and Wednesday.

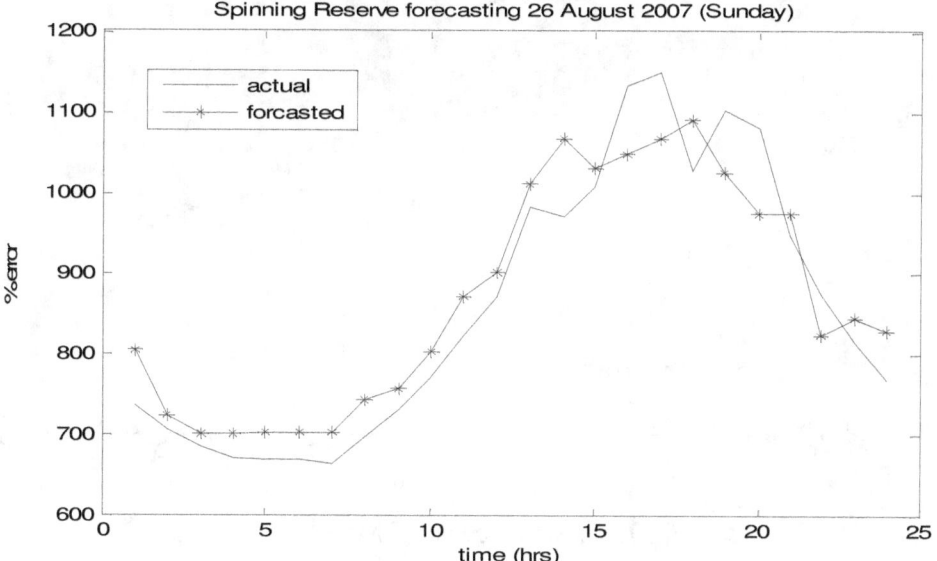

Fig 7.23: SR forecast of Sunday (summer)

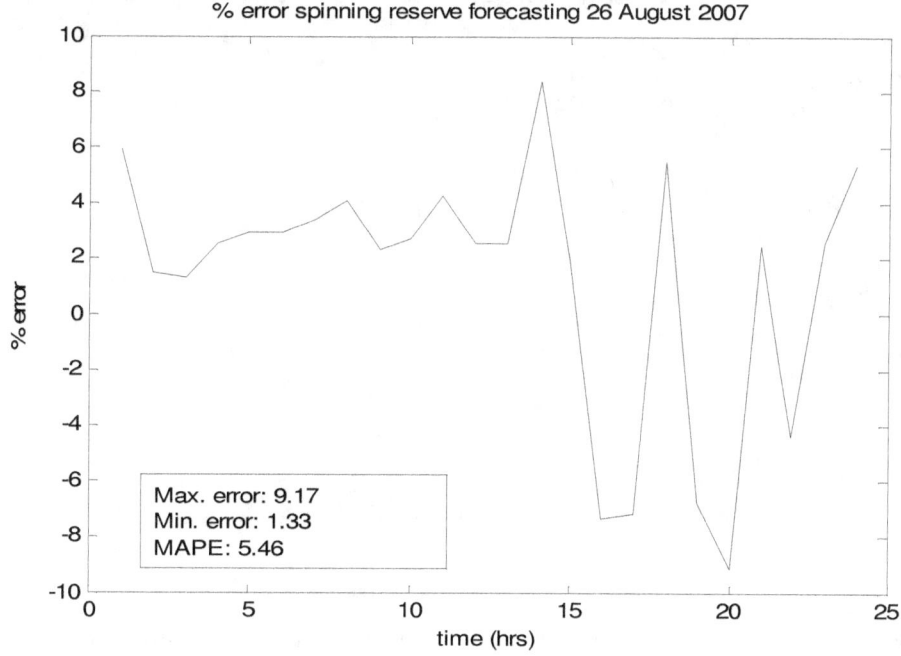

Fig 7.24: Error in SR forecast of Sunday (summer)

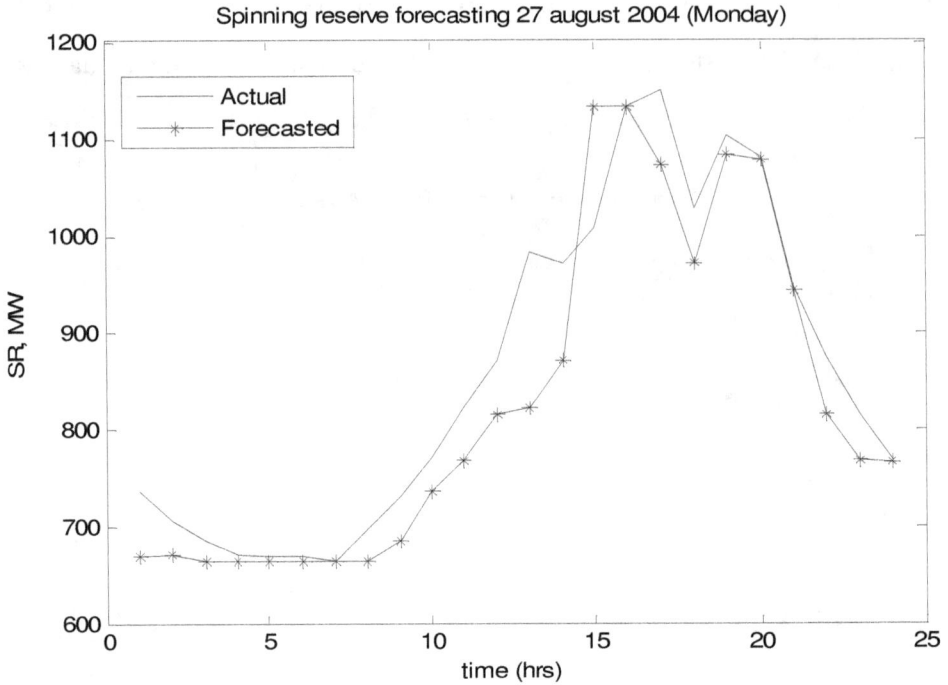

Fig 7.25: SR forecast of Monday (summer)

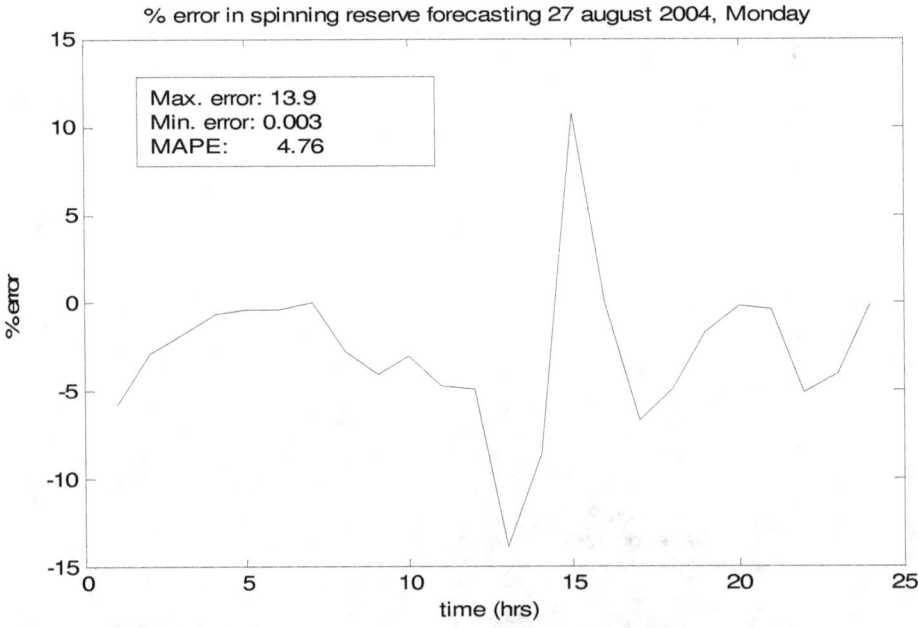

Fig 7.26: Error in SR forecast of Monday (summer)

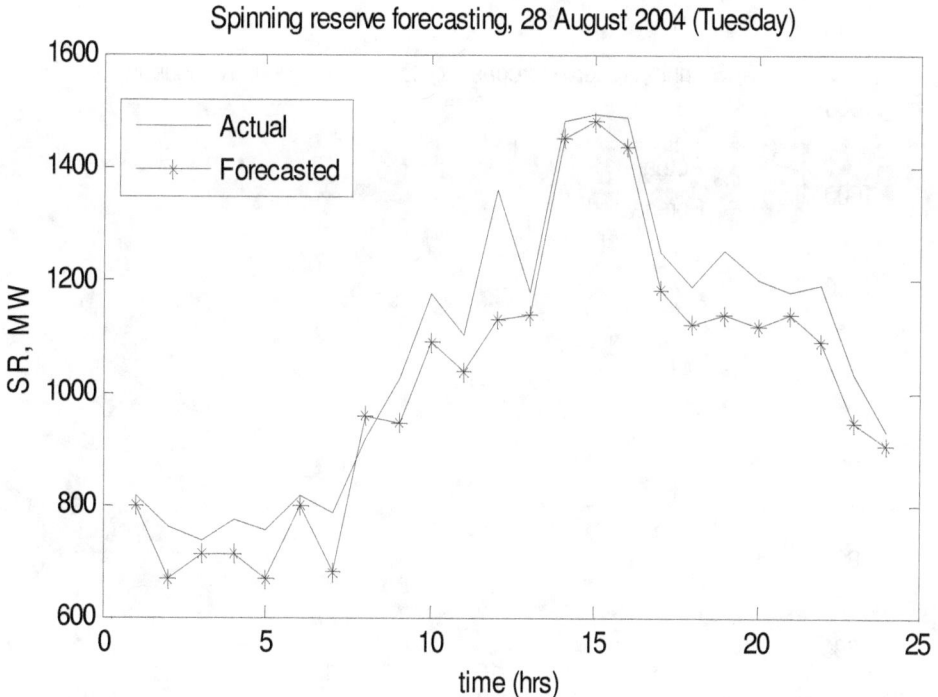

Fig 7.27 SR forecast of Tuesday (summer)

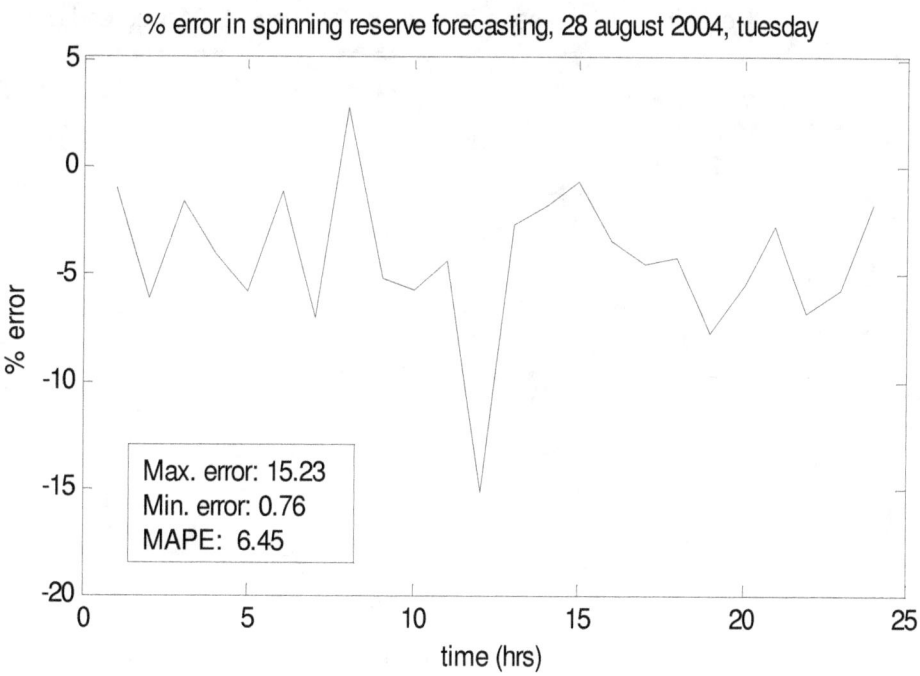

Fig 7.28 Error in SR forecast of Tuesday (summer)

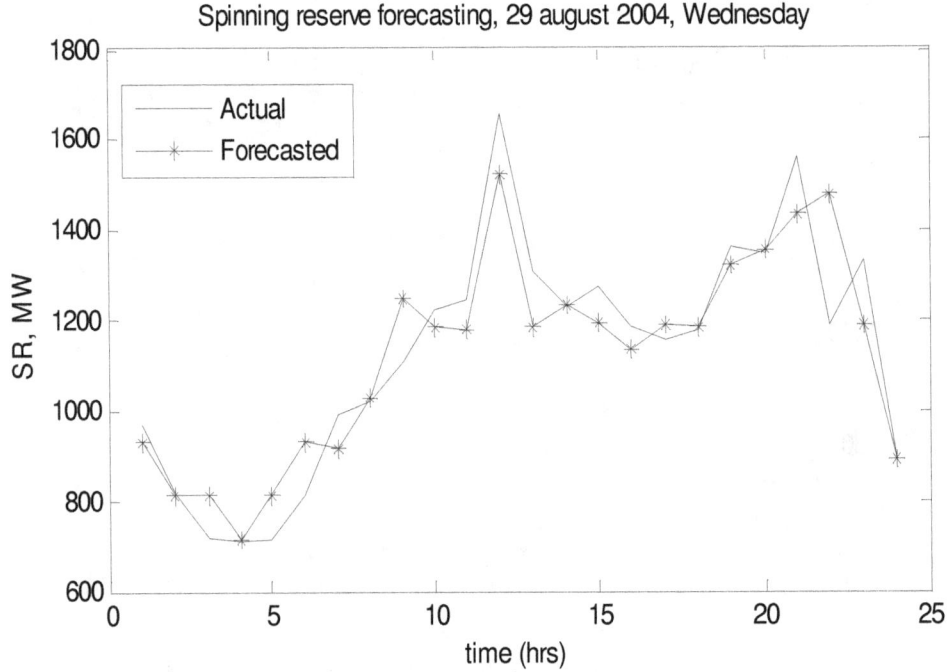

Fig 7.29 SR forecast of Wednesday (summer)

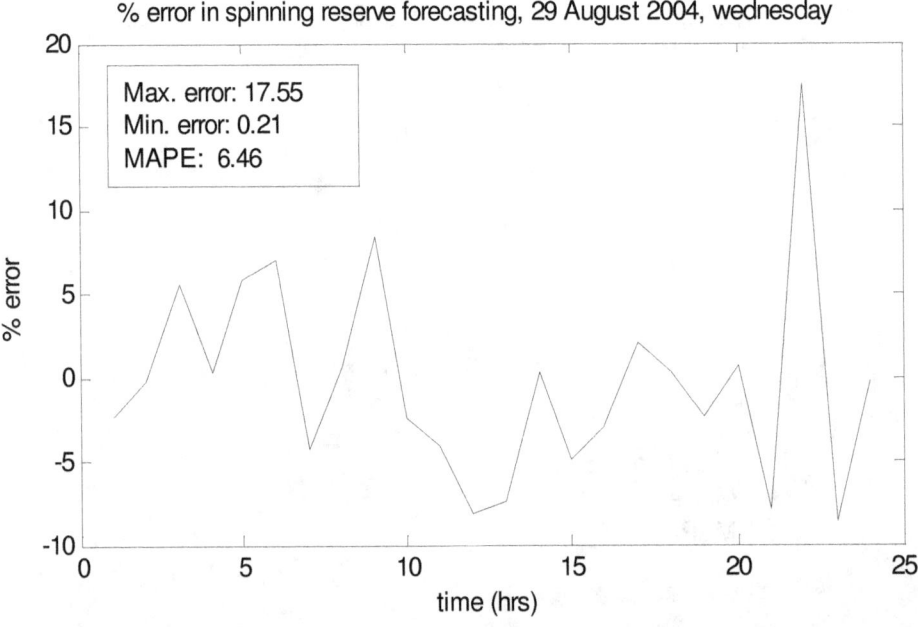

Fig 7.30 Error in SR forecast of Wednesday (summer)

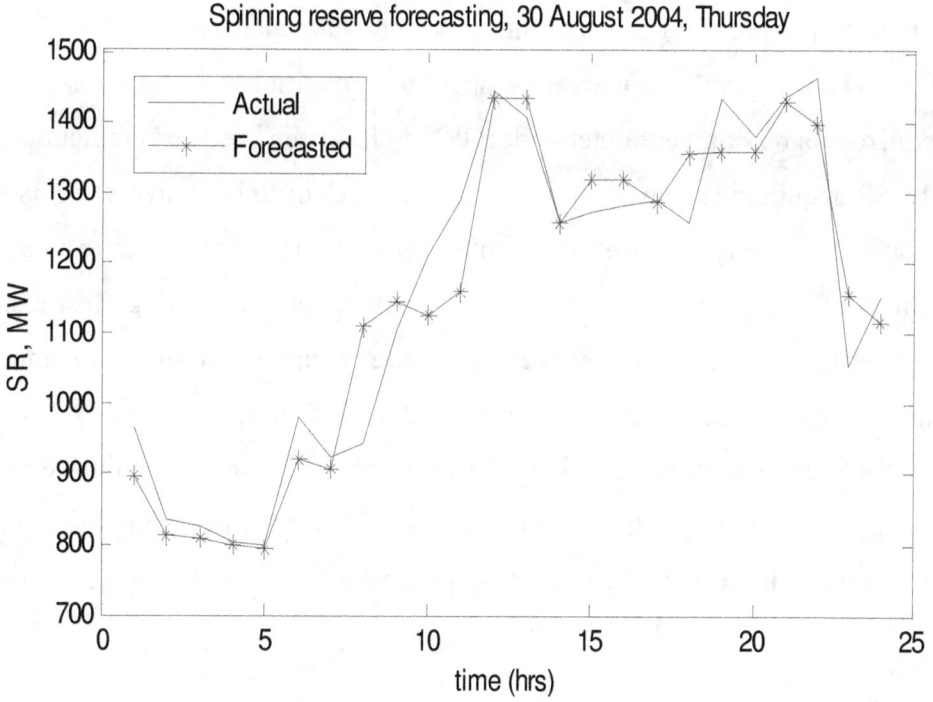

Fig 7.31 SR forecast of Thursday (summer)

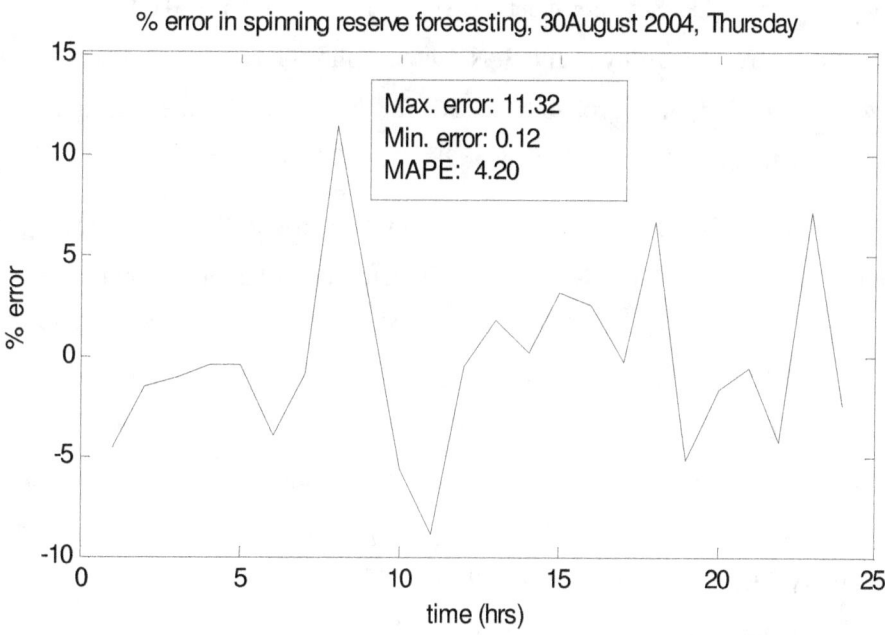

Fig 7.32 Error in SR forecast of Thursday (summer)

7.6   Summary

The spinning reserve (SR) is one of the most important ancillary services (AS) procured by the CAISO to respond to emergency operating conditions and unforeseen load swings. Generalized regression neural network (GRNN) has been used for prediction of day-ahead hourly SR requirements in electricity market of California. Forecasting is done for all weekdays of summer as well as winter seasons. The presented results confirm the usefulness of GRNN model for hourly SR requirements forecasting. Test results obtained shows that forecasting for winter season is more accurate than that for summer. The tests results obtained by the GRNN and modified GRNN are compared with the forecasted hourly SR requirements published by CAISO and other methods published in literatures. The results of comparison show that GRNN is good in SR forecasting but modified GRNN method is better in SR forecasting and as effective as other methods like wavelet-ANN.

References

[1] Jinxiang Zhu, Gary Jordan and Satoru Ihara, "The market for spinning reserve and its impacts on energy prices", IEEE PES winter meeting, Vol. 2, Jan. 2000, pp. 1202-1207.

[2] M. M. Tripathi, K. G. Upadhyay and S. N. Singh, "Short-Term Load Forecasting using Generalized Regression and Probabilistic Neural Networks in the Electricity Market", The Electricity Journal, Vol. 21, No. 9, November 2008, pp. 24-34.

[3] M. M. Tripathi, K. G. Upadhyay and S. N. Singh, "Electricity Price Forecasting using General Regression Neural network (GRNN) for PJM Electricity Market", International Review of Modeling and Simulation (IREMOS), Volume 1, No. 2, December 2008, pp 318-324.

[4]A report on "Ancillary services capacity settlement in CAISO controlled grid," Available at http://www.caiso.com.

[5] N. M. Pindoriya, S. N. Singh and S. K. Singh, "Forecasting the Day-Ahead Spinning Reserve Requirement in Competitive Electricity Market", Conversion and Delivery of Electrical Energy in the 21st Century, PES General meeting, 20-24 July 2008, pp. 1 – 8.

[6] Open access same time information system (OASIS) of California electricity market. Available at http://oasis.caiso.com/.

[7] A. J. Rocha Reis and A. P. Alves da Silva, " Feature extraction via multiresolutionanalysis for short-term load forecasting", IEEE Trans. Power Systems, Vol. 20, No. 1, Feb. 2005, pp. 189–198.

Appendix – A

Electronics References and Information

The websites from where historical load, price, spinning reserve and weather data were collected are listed below.

1. **PJM electricity market**

   http://www.pjm.com/markets-and-operations.aspx

2. **National Electricity Market Management Company Limited**

   http://www.nemmco.com.au

3. **Australian Energy market Operator**

   http://www.aemo.com.au

4. **California Independent System Operator Corporation Ancillary Services Market**

   http://www.caiso.com

5. **Bureau of Meteorology, Govt. of Australia**

   http://www.bom.gov.au/weather/vic

6. **National Climatic Data center, USA**

   http://www.ncdc.noaa.gov/oa/ncdc.html

ACKNOWLEDGEMENT

I am grateful to Prof. S. N. Singh, Prof. K. G. Upadhyay, Dr. Neeraj Pandey and Sh. Anil Kumar Pandey for their contribution in the book by providing valuable suggestions and guidance without which it was not possible to write this book. I also acknowledge the help of PJM Electricity Market, NEMMCO Electricity Market and California Electricity Market from which web sites from were historical data has been used to show that how forecasting based on ANN is useful for electricity market. We also acknowledge the Delhi Technological University, Delhi, India for providing Hardware and Software support for carrying out the work.